Instrumentation Handbook for Biomedical Engineers

Instrumentation Handbook for Biomedical Engineers

Mesut Sahin
Howard Fidel
Raquel Perez-Castillejos

CRC Press is an imprint of the
Taylor & Francis Group, an **informa** business

First edition published 2021
by CRC Press
6000 Broken Sound Parkway NW, Suite 300, Boca Raton, FL 33487-2742

and by CRC Press
2 Park Square, Milton Park, Abingdon, Oxon, OX14 4RN

First issued in paperback 2022

Visit the Taylor & Francis Web site at
http://www.taylorandfrancis.com

and the CRC Press Web site at
http://www.crcpress.com

ISBN: 978-0-367-56668-5 (pbk)
ISBN: 978-1-4665-0466-0 (hbk)
ISBN: 978-0-429-19398-9 (ebk)

DOI: 10.1201/9780429193989

Typeset in Minion
by Deanta Global Publishing Services, Chennai, India

Visit the companion eResources: www.routledge.com/9781466504660

Contents

Foreword

TEACHING BIO-INSTRUMENTATION CAN BE both rewarding and challenging. It has been a great opportunity for me to teach this material as I continued to learn more about this fascinating topic of blending instrumentation with the body. But it has been difficult as it requires integration of knowledge of both physiology and engineering. Bioinstrumentation can be taught from different angles but this book has chosen a road less travelled by focusing of a particular physiological need and providing engineering solutions. This pedagological method makes it easier to get into the topic of each chapter and to bring engineering knowledge to bear on the issue. Each topic has a laboratory associated with detailed instructions for hands-on learning experience. I highly recommend this book as the combination of didactic material with laboratory protocols will make learning bioinstrumentation easy and entertaining.

Dominique M. Durand, PhD
Distinguished University Professor
Department of Biomedical Engineering
Case Western Reserve University
Cleveland, Ohio

Preface

IT HAS BEEN MY passion ever since I started teaching at my first post as a faculty to write a book that takes students through practical hands-on projects, the best way of learning, and makes bioelectronics a joyful experience. My primary objective by putting this book together is to help students gain confidence in building bioinstrumentation circuits and understand that it is not a challenging task that they should fear.

This book is unique in its content with laboratory exercises that are carefully chosen for biomedical engineering undergraduate students, and clearly fills a void among the textbooks available for teaching. The content of the book took many years to develop and refine as a spin-off project while I have been teaching laboratory courses using this material at the New Jersey Institute of Technology for junior/senior level undergraduates over the past 15 years. Each circuit has been built and tested by multiple cohorts of students each year during classes. Data generated by students are included in most chapters wherever appropriate. I believe seeing signals collected by their peers will be very appealing to the students using this book, although some plots may be prepared haphazardly as a part of their laboratory report.

I share the authorship of this book with two colleagues who contributed to the book with their unique expertise on the matter. Howard Fidel is truly a superb biomedical engineer and entrepreneur with his career, spanning 40 years, dedicated to designing circuits and devices. Howard reviewed the technical details of the circuits in the book, made comments, and contributed an Introductory chapter and a laboratory exercise (Chapter 13) to the book. Dr. Raquel Perez-Castillejos edited the text and contributed excellent drawings and background information to multiple chapters. I feel that the expertise of both co-authors immensely improved the quality of the content and the presentation.

We thank Dr. Joel Schesser for contributing the description on the usage of oscilloscopes to the Introduction.

Mesut Sahin, PhD
Professor of Biomedical Engineering
New Jersey Institute of Technology

About the Authors

Mesut Sahin earned his B.S. degree in electrical engineering from Istanbul Technical University in 1986. After graduation, he worked for a telecommunication company, Teletas A.S., in Istanbul in hardware and software development of phone exchanges until 1990. He earned the M.S. degree in 1993 and a Ph.D. degree in 1998, both in biomedical engineering, particularly in the field of neural engineering, from Case Western Reserve University, Cleveland, Ohio. After post-doctoral training at the same institute, he joined Louisiana Tech University as an Assistant Professor in 2001. He has been on the faculty of Biomedical Engineering at New Jersey Institute of Technology, Newark, New Jersey since 2005, and currently is a Full Professor, where he teaches bioinstrumentation and neural engineering courses. His research interests are mainly in neural modulation and development of novel neural prosthetic approaches. He has authored more than 90 peer-reviewed publications. Dr Sahin is an Associate Editor of IEEE Transactions on Biomedical Circuits and Systems and a Senior Member of IEEE/EMBS.

Howard Fidel served as Vice President of Technology for IREX/Johnson and Johnson Ultrasound, where he developed the market leading Meridian Cardiology system. After leaving Johnson and Johnson in 1986, he founded Universal Sonics Corporation, as a contract engineering company and a manufacturer of OEM Medical Devices supporting the Ultrasound and Medical Device industry. Universal Sonics' customer base included many well-known clients, including ATL, Acuson, Biosound, and many others. After the acquisition of Universal Sonics by US Surgical, Mr. Fidel left to become Chief Operating Officer of Stern Ultrasound, a start-up company that was in the process of developing a mid-market ultrasound system. Later he functioned as C.T.O. and C.O.O of 3G Ultrasound.

In 2017, Mr. Fidel became an adjunct Professor at the New Jersey Institute of Technology teaching biomedical engineering. Mr. Fidel earned a BE degree with High Honors from Stevens Institute of Technology in 1972, an MS in Bioengineering from the University of Connecticut in 1974, and an MBA with Distinction from Pace University in 1984. He holds eight patents in the ultrasound field. In 2016, he was inducted into the New Jersey Inventors Hall of Fame. He currently lives in Tarrytown, NY and grew up in Brooklyn, NY. He is married to Professor Marlene Brandt Fidel, and has a daughter, Rivka Fidel, Ph.D.

Raquel Perez-Castillejos earned her B.S. degree in telecommunications engineering with a specialization in Microelectronics from the Polytechnic University of Catalonia in 1996. She earned her Ph.D. from the Institute of Microelectronics of Barcelona in 2003, followed by a post-doctoral stay in 2003–2004 at the University of São Paulo and postdoctoral training at Harvard University, Department of Chemistry and Chemical Biology, from 2004 to 2008. She joined the faculty of the New Jersey Institute of Technology from 2008 to 2016 and is currently an Independent Consultant in Biomedical Devices specialized in applications of microelectronics and microfluidics for cell biology and biochemical analyses. Dr. Raquel Perez-Castillejos has authored and co-authored 29 peer-reviewed papers, 5 patents, and more than 50 conference proceedings abstracts.

Abbreviations

AC	Alternating Current
DAQ	Data Acquisition Board
DC	Direct Current
ECG	Electrocardiography or Electrocardiogram
EKG	Electrocardiography or Electrocardiogram
HF	High Frequency
HRV	Heart Rate Variability
IR	Infra-Red light
LED	Light-Emitting Diode
LF	Low Frequency
Op-Amp	Operational Amplifier
PG	Plethysmograph
PPG	Photoplethysmograph
RSA	Respiratory Sinus Arrhythmia
SA node	Sinoatrial Node
ULF	Ultra-Low Frequency
VLF	Very Low Frequency
V_s	Supply Voltage
V_{col}	Collector Voltage; voltage at the collector terminal of a transistor

Introduction

I.1 ORGANIZATION

This book is organized into Studios, or experiments, that the student or students will perform in conjunction with advanced biomedical engineering courses in the junior or senior year of an undergraduate program. Each Studio is independent and need not be done in the order presented. The instructor should determine which Studios are most appropriate for the goals of the particular course. The instructions and a list of required material is presented in each Studio. A summary of all the required material is provided in the Appendices to help simplify material acquisition. In the next section of this Introduction we will review the equipment used in the Studios and the methods used to assemble and test the Studios.

I.2 EQUIPMENT

The minimum equipment required for these Studios is:

1. Oscilloscope.
2. Digital multimeter.
3. Dual power supply.
4. Signal generator.
5. Physiological amplifier.
6. Plug-in protoboard.
7. Miscellaneous electronic components (see appendices).

8. Miscellaneous cables such as:
   a. Oscilloscope probes.
   b. BNC to clip lead cables.
   c. Banana jack to banana jack cables.
   d. Insulated solid wire 24 AWG for protoboard or prebuilt jumper wires.
9. Miscellaneous hand tools:
   a. Wire strippers.
   b. Side cutters.
   c. Needle nose pliers.
10. Disposable ECG electrodes and conductive gel.

I.3 OSCILLOSCOPE*

An oscilloscope (scope for short) displays a voltage waveform versus time and has the following components:

1. A screen to display a waveform.
2. Input jacks for connecting the signal to be displayed.
3. Dials to control how the signal will be displayed.

The screen is divided into a two-dimensional grid (or axes or scale); say a 10×10 grid. In this instance, the vertical grid is divided up into ten (major) divisions and the horizontal grid is divided into ten major divisions. To improve the precision, each of these divisions is further broken up into five minor divisions.

The horizontal axis (X-axis) represents time and the vertical axis (Y-axis) represents voltage. The scope displays (also called a signal trace or trace) the input signal voltage along the vertical (or Y-axis) while an internally generated signal (called the horizontal sweep ramp) is simultaneously produced along the X-axis creating a two-dimensional time trace of the input signal. So if the scope is set to 1 volt/major vertical division

* *Oscilloscope Operation and Basic Measurements, kindly donated by Dr J. Schesser of New Jersey Institute of Technology*

and 0.5 seconds/major horizontal division, then a signal that occupied two vertical divisions would be 2 V peak-to-peak amplitude. If the signal was a repetitive waveform, such as a sine wave, and if a complete cycle of the sine wave occupied two horizontal boxes, then one cycle would take 1 second and be a 1 Hz sine wave. Figure I.2 shows a sine wave with amplitude of 1 volt and a frequency of 1 Hz. (Here we have a 4 × 10 grid for the display.)

I.3.1 Signal Inputs

There is at least one set of connections on each oscilloscope for connecting the external signal to be displayed. Modern scopes can display two or more signals at a time and, therefore, would have a set of jacks for each signal to be displayed. These are sometimes called **Y-Inputs**. Sometimes there are other jacks to connect signals which are used as references but may not be used to display. These inputs are called **Trigger Inputs**. As part of the **Y-Input** jacks there may be a switch to directly connect (**DC**) or capacitively connect (**AC**) the signal to the scope. The latter passes any DC component of the signal while the latter filters out the DC.

There are several controls on the scope, which include: the vertical grid (or scale) control (**Volts/Div**), vertical position control, the horizontal scale control (**Timebase**), intensity control, **Trigger Level**, **Trigger Source**, etc. There are vertical controls for each **Y-Input** supported by the scope. The intensity control controls the brightness of the trace and the vertical position control is used to set the zero voltage value of the signal along the Y-axis (e.g., at the center of the grid).

The basic scope controls are the vertical (**Volts/Div**) and horizontal (**Timebase**) controls. The vertical scale control is used to set how one reads the voltage values from scope's Y-axis grid. This is called the **Volts/Div**. Looking at Figure I.1 again, we see a sine wave with amplitude of +/−1 volt and the **Volts/Div** is set to 0.5 volts/division, so the sine wave is four boxes from peak to peak (p-p). (Here we have a 4 × 10 grid for the display.)

In Figure I.2 the same sine wave is displayed; however, the **Volts/Div** is set to 1 volt/division.

If you set the **Volts/Div** too low, you will clip the signal. Figure I.3 shows the same sine wave with a **Volts/Div** of 250 millivolts/division.

Likewise, setting it too high, and you will not find the signal. Figure I.4 shows the same sine wave with a **Volts/Div** of 50 volts/division.

Therefore, knowing the approximate voltage maximums and minimums of the input signal should be the guiding factor for choosing an appropriate value for the **Volts/Div**.

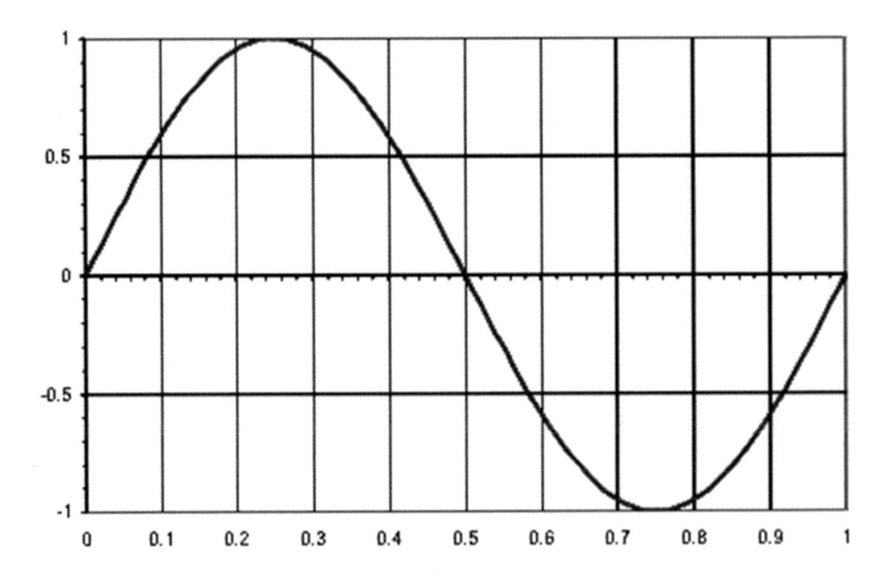

FIGURE I.1 A +/−1 volt sine wave with a frequency of 1 Hz with a vertical scale of 0.5 volts/div.

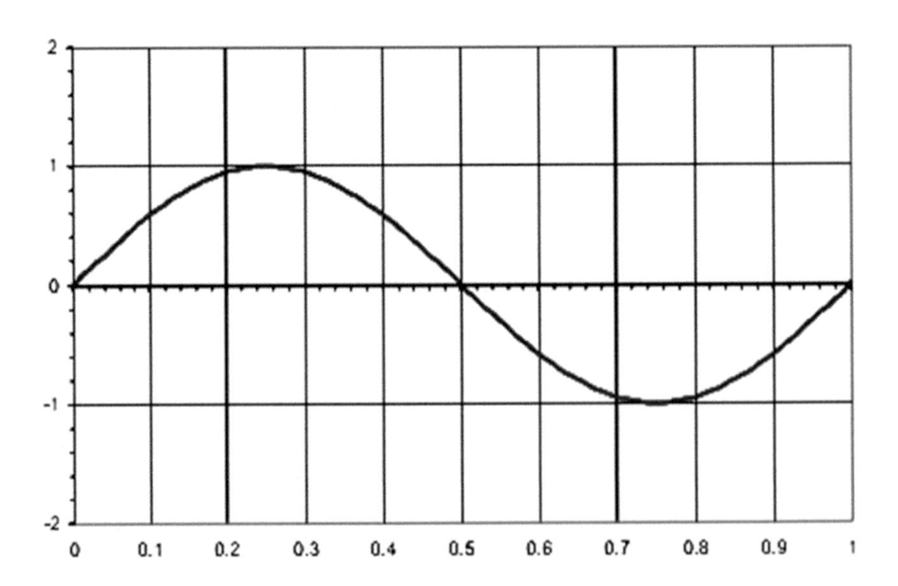

FIGURE I.2 A +/−1 volt sine wave with a frequency of 1 Hz with a vertical scale of 1.0 volts/div.

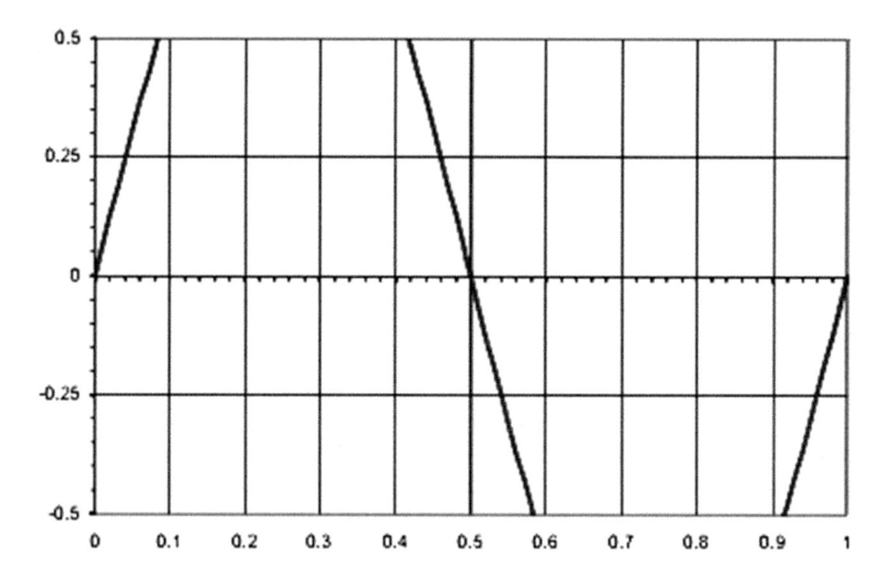

FIGURE I.3 A +/−1 volt sine wave with a frequency of 1 Hz with a vertical scale of 0.25 volts/div.

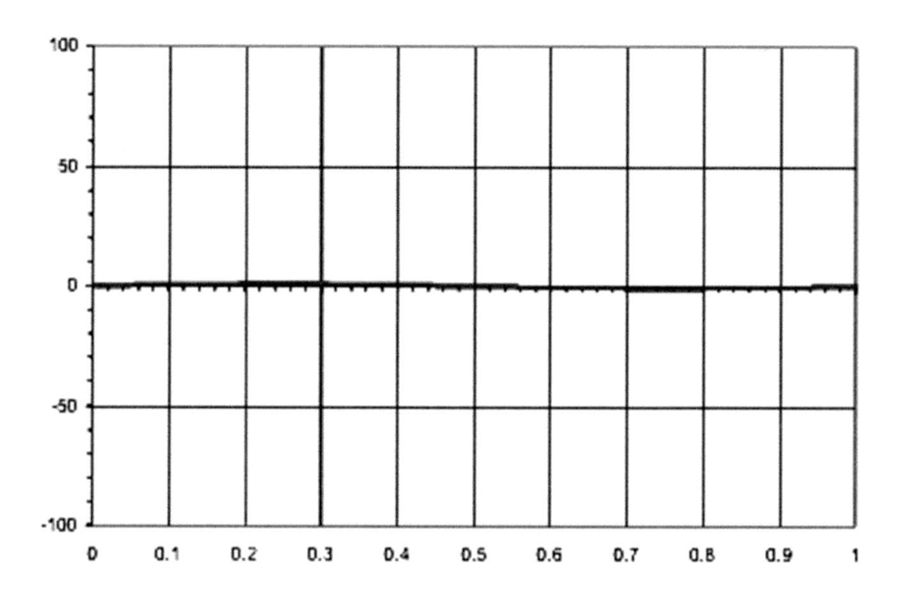

FIGURE I.4 A +/−1 volt sine wave with a frequency of 1 Hz with a vertical scale of 50 volts/div.

If we assume that in all the above screen captures the **Timebase** controls are set to 100 milliseconds/division and there are ten divisions on the horizontal axis, then one cycle takes ten squares and is 1 second long. Setting the **Timebase** to 200 milliseconds/division × 10 divisions = 2 seconds will yield a display of two cycles as shown in Figure I.5.

Therefore, increasing the **Timebase** will display more cycles of a periodic signal. Increasing too much will clutter the display. Conversely, reducing the **Timebase,** fewer cycles will be displayed. In the following figure, the **Timebase** is set to 50 milliseconds/division and one half of a cycle is displayed. Reducing is too much may display a useless fragment of a cycle (Figure I.6).

Figure I.7 shows the trace of a square wave with a frequency of 4 Hz. Therefore, knowing the approximate maximum frequency of the input signal is the guiding factor for choosing an appropriate value for the **Timebase.** Recall that the inverse of the maximum frequency of a periodic signal will yield the (time) **period** of one cycle. Therefore, the **Timebase** should be calculated by taking the **period** of the signal and dividing it by the number of horizontal X-axis divisions times the desired number of cycles to be displayed.

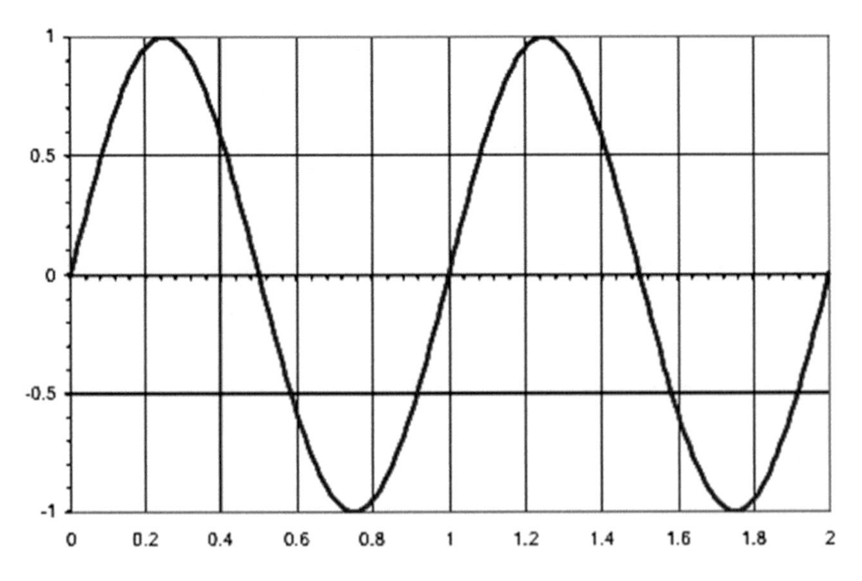

FIGURE I.5 A +/−1 volt sine wave with a frequency of 1 Hz with a vertical scale of 0.5 volts/div and a horizontal scale of 0.2 sec/div.

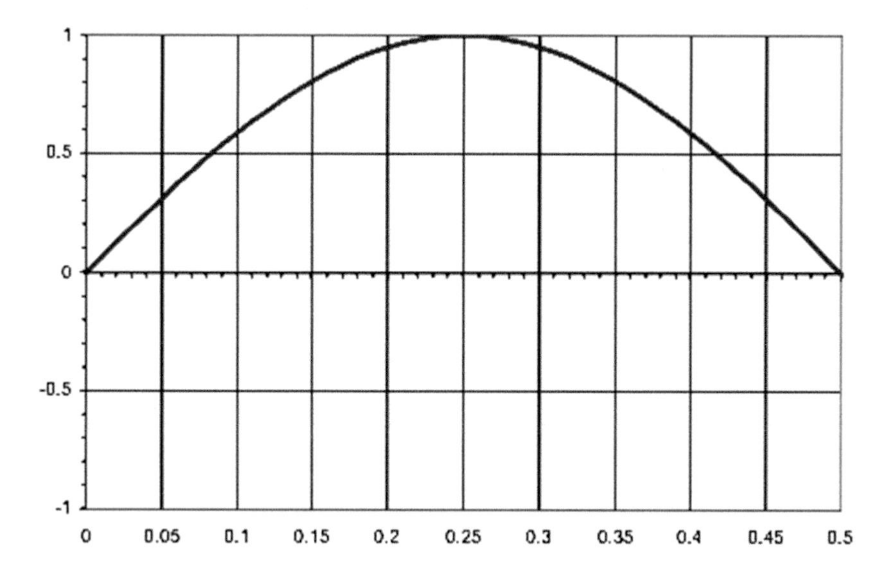

FIGURE I.6 A +/−1 volt sine wave with a frequency of 1 Hz with a vertical scale of 0.5 volts/div and a horizontal scale of 0.05 sec/div.

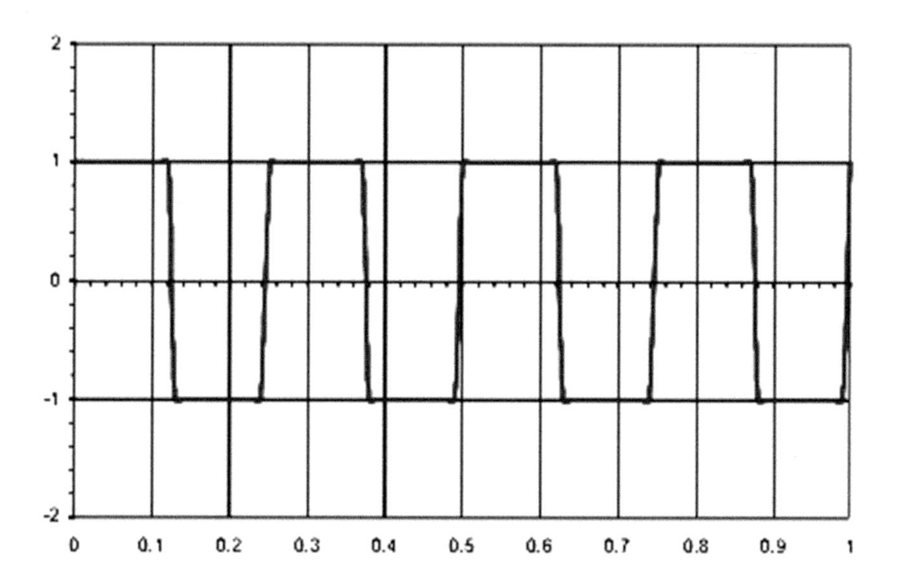

FIGURE I.7 A +/−1 volt square wave with a frequency of 4 Hz with a vertical scale of 1.0 volts/div and a horizontal scale of 0.1 sec/div.

I.3.2 Trigger Level and Trigger Source

The **Trigger** is used to determine where on the input signal to start the horizontal sweep of the oscilloscope. Looking at our sinusoidal wave signal, we can see that the amplitude is +/–1 volt. If the trigger level is set to any value between –1 and +1 volt the sweep will be triggered to start at that voltage. You may select a positive trigger **Slope**, which would mean that the display will start on the positive going edge in the waveform. If you select a negative slope, the sweep will trigger at the selected voltage on the negative edge. Most oscilloscopes indicate the trigger voltage on the display numerically and with an arrow to the right of the screen at the trigger amplitude. If the trigger is set higher than +1 volt or lower than –1 volt, and the trigger mode is set to **Normal**, the display will not trigger or update. If it is set to **Auto** mode, the display will sweep, but in most cases the waveform will not appear to be stationary because it is no longer triggered. You may select any input channel as the trigger source, or an external signal. Some oscilloscopes have special triggering modes, such as line voltage (the AC input to the oscilloscope).

I.4 DIGITAL MULTIMETER

A digital multimeter is a device with a numeric LCD. Depending on the precision of the meter, it has 2.5, 3.5 or more digits. The 0.5 digit implies that the first digit is either 1 or absent. No matter what scale is selected the largest number that can be displayed on that scale is 1.99 depending on how many digits the display is. If the meter is set to the 2 volts scale, then the maximum voltage that can be displayed is 1.99 volts, if it goes higher, the meter will over-range and display 1 and no additional digits. A basic multimeter will measure volts, amperes, and ohms. The maximum reading that can be displayed on the scale can be selected with a rotary knob. Many times, the off position is on this knob, if not, there is a separate on-off switch. Some meters have an auto off feature to extend the battery life. If the meter turns off, just change the scale and it will wake up. All meters have an input, which are usually banana jacks. One lead is labeled COM, which is the negative lead, and the black lead is inserted there. The red lead is plugged into the V or Ω jack. On some meters current is also measured at this same input jack. On others, current has one or two additional input jacks. Most meters have a 10 ampere current input jack that is separate from the regular current measurement jack and is used for currents higher than 200 mA. The red lead would be

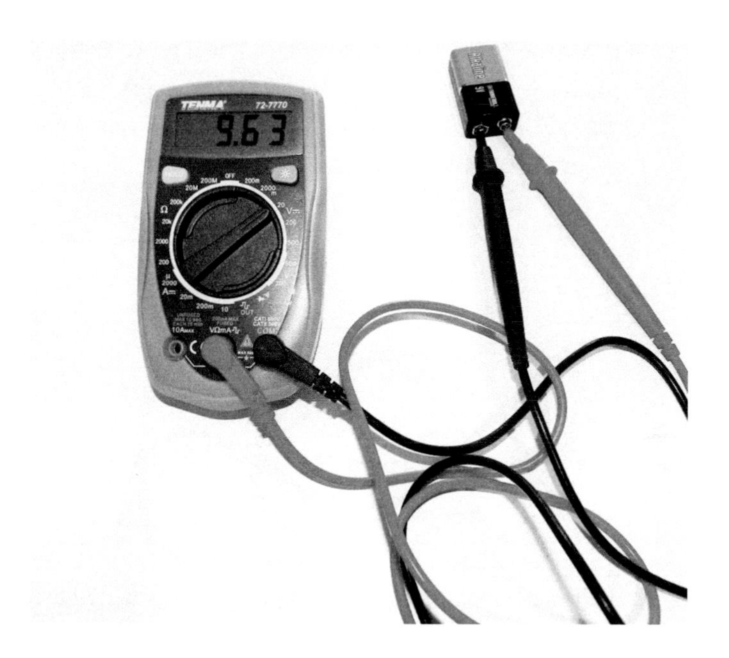

FIGURE I.8 A multimeter measuring a battery voltage.

moved there to measure current. To measure voltage, set the scale to an appropriate range for what you are measuring. The highest voltage we use in this text is 15 volts, so you normally would not need to use a scale higher than 20 volts. All our measurements will be DC. See Figure I.8. To make AC measurements, you need to select the AC voltage range, not the DC range. The AC scale is usually marked with a ~ next to V sign (~V). The DC range will be marked with a V followed by a dash (V-). If you reverse the leads and put the black lead on a positive voltage, and the red lead to ground, then the voltage will read negative, even though it is positive. If you do this, you can mentally make the correction, re-measuring is not required. You can use the ohmmeter to measure resistance of the resistors you use before you put them in the circuit. Once you assemble your circuit, you can use the ohmmeter to check for shorts or opens, but you should be aware that resistance measurements of a particular resistor may not be accurate since other resistive components maybe connected to it in the circuit. Many ohmmeters have *continuity checkers* and when you put the leads across a low impedance, the meter buzzes, like saying, "you are buzzing out the circuit."

To measure current, the leads of the meter are put in series with the component that the current of which is being measured. So, if you

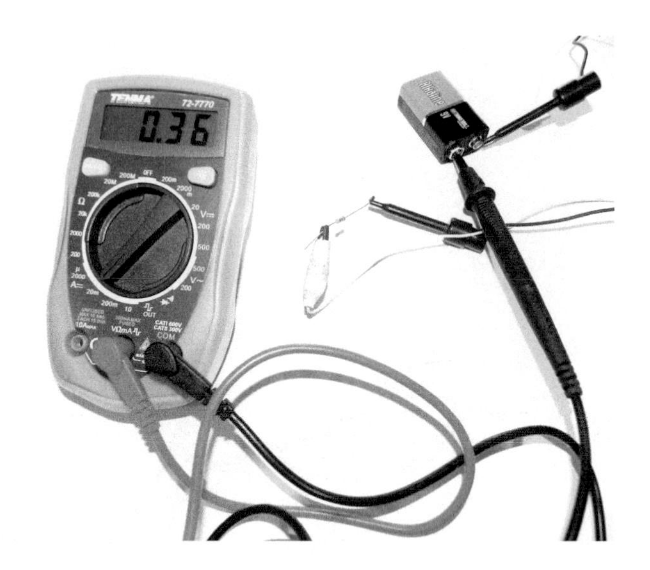

FIGURE I.9 A multimeter measuring current through a resistor. (Note resistor is in series with the battery.)

wanted to measure the current through a resistor in your circuit you should lift one lead of the resistor out of the circuit, and connect the meter between the resistor lead and the point where the resistor was plugged in (Figure I.9).

Some multimeters have extra features, such as the ability to measure capacitance, frequency, transistor check and diode check.

I.5 DUAL POWER SUPPLY

A dual power supply Figure I.10 supplies two output voltages. The two outputs can usually be adjusted independently or slaved. In the slaved mode the second supply is set to the same voltage as the first supply as the knob is dialed. This is useful for making dual voltage supplies, such as +/−15 volts. The supplies usually also have a current limit adjustment. The power supply will have an adjustable fold back current limit. If the current set is exceeded, the output voltage will be reduced to keep the current at or below the set limit. In some supplies, when this happens the load will need to be disconnected and reconnected to allow the voltage to return to the set point after the overload condition is removed. Sometimes if the current knob is set too low, this condition may occur when the circuit is functioning normally.

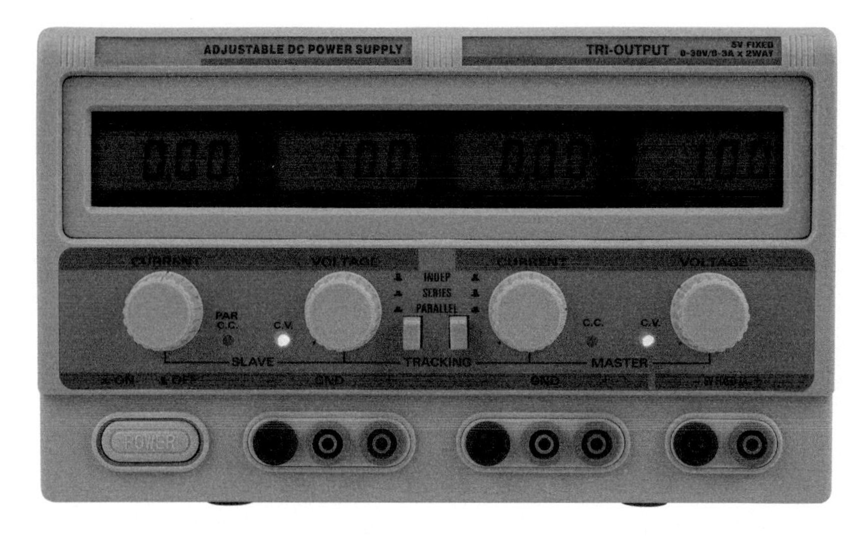

FIGURE I.10 A typical triple output power supply. The third output (shown in the right here) is usually fixed, or adjustable over a limited range.

I.6 FUNCTION GENERATOR

A function or signal generator is a device that produces different types of periodic waveforms. The common output waveform types are:

- Sine waves.
- Square waves.
- Sawtooth waves.
- Triangle waves.

The function generator Figure I.11 gives the user the ability to select the function type, the frequency and amplitude of the function. The frequency is set with a range control and a knob in the case of the unit pictured below. Other generators may have a keypad entry to directly enter the frequency. The amplitude can be adjusted with a knob, and also in this case a –20 dB button that reduces the output voltage by a factor of ten. Normally the function waveform is centered around 0 volts. A DC offset can be added to shift this up or down. The generator may have more than one output. Here there are two, one for the function selected, and one for logic drive labeled TTL/CMOS. This allows logic circuitry to be driven

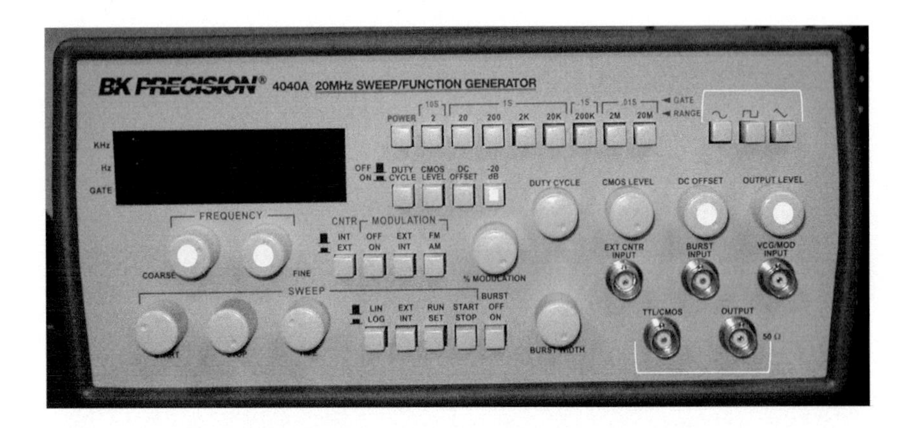

FIGURE I.11 Function generator.

by a logic level voltage, usually starting at ground and going positive. The signal generator shown below also has modulation function capability and frequency sweep capabilities. The four knobs that control the functions discussed here are marked with markers in the picture.

I.7 PHYSIOLOGICAL AMPLIFIER

A physiological amplifier is a special amplifier designed to meet the needs of amplifying physiological signals. This kind of amplifiers have a built-in calibrator that allows for easy calibration of the output waveform amplitude. They have adjustable low-pass and high-pass filters that can be set to optimally pass different types of physiological signals, and they have a gain control range that covers what is needed to amplify these signals and make them readily visible on a display. A Grass brand amplifier is shown in Figure I.12.

I.8 PROTOBOARD

The protoboard (also called breadboards) is a device designed to interconnect electronic components without the need for soldering, which greatly simplifies the process of prototyping electrical circuits for evaluation. These boards have arrays of interconnected ports (holes) where the electronic components can be inserted and electrically connected without using solder. It is important to note that protoboards providing a two-dimensional array of through holes (Figure I.13) are not intended for very high frequencies (> 1 MHz) or high voltages. In Figure I.13 the horizontal groove down the middle is the location to plug integrated circuits (ICs) of the DIP (dual inline pin) type with 0.3” lead spacing. Above and below the

FIGURE I.12 Physiological amplifier.

FIGURE I.13 Protoboard.

gap are vertically arranged in groups of five ports or holes. Because the five holes in one vertical group of five are connected together, wires and component leads plugged into any of the five holes of the vertical group will be connected together. One of those vertical groups is marked by a black line in the picture. Along the top and bottom are two rows that are reserved for power supply and ground distribution. Holes in each of these two rows are connected together from each end to the middle. Notice that in Figure I.12 the two halves in each of the four horizontal rows are connected by means of a short green wire.

I.9 TROUBLESHOOTING

Sometimes after assembling a circuit, it does not function properly or at all. In these cases, it is important to be able to determine what is wrong and fix the problem. Usually the problem is an assembly error. So, the first step is to double check your work and make sure the circuit assembled matches the schematic. Common errors of this type are using a wrong component; you may have misread the color code of a resistor for example and not checked it first with an ohmmeter before putting it into the board. Another common problem is plugging a lead into the wrong column either to the left or the right of the correct one. Sometimes the power is not wired up correctly, or it is not getting to the voltage input pins of the integrated circuits (ICs). These kinds of issues can be checked easily with a voltmeter connected directly to the ICs power pins. Sometimes the component used is defective. Components like diodes and transistors can be removed from the circuit and checked with an ohmmeter. A diode will exhibit a low resistance in one direction and a high resistance when the leads are reversed. Make sure you plugged in the diode correctly, with the cathode (where the diode package shows a minus sign) at the lower voltage column. Although we do not use any discrete transistors in this book, they can also be easily tested with a multimeter by checking the emitter-base and base-collector junctions in a similar fashion, or if your meter has one, the h_{fe} (current gain) test will indicate if the transistor's gain is high enough. ICs are more difficult to test. If your circuit passes all the tests mentioned above, then the problem may lie with your IC. First make sure you located pin #1 properly. If it is, then the fastest solution to this issue is to substitute the possibly faulty IC with a new one and see if it corrects the problem. Capacitors can be challenging to test as well. If your meter can measure capacitance, that is the best method. Otherwise, if the capacitor value is large – i.e., in the microfarad range – it is usually possible to see the capacitor charge up when connected to the ohmmeter leads. The resistance initially is low, and then increases to that of an open circuit. It is recommended to use a 200 kΩ scale to do this test. With electrolytic capacitors, it is important to have the capacitor correctly connected in terms of polarity, with the black stripe (minus sign) of the capacitor package indicating the negative lead.

STUDIO 1

Body Thermometer Using a Wheatstone Bridge and the Projection Method

S1.1 LEARNING OBJECTIVES

The objectives of the studio are for the students to:

1. Learn the advantages of using a Wheatstone Bridge.
2. Become familiar with the projection method based on exponential function.
3. Understand the operation of a thermistor.

S1.2 BACKGROUND

A Wheatstone Bridge is a method of using two voltage dividers to gain increased sensitivity to small resistive value changes. The circuit for one voltage divider is shown in Figure 1.1.

The voltage at B with respect to C can be calculated as follows:

$$V_B = V_{BT1} R_2 / (R_1 + R_2)$$

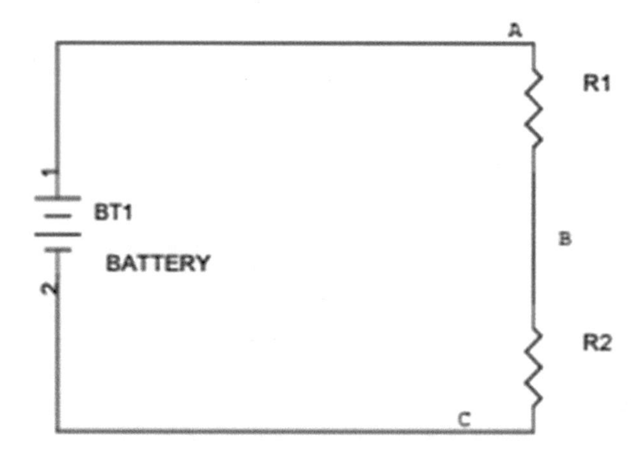

FIGURE 1.1 Voltage divider.

If we set R_1 equal to R_2 then:

$$V_B = V_{BT1} / 2$$

If for example BT1 is 9 volts, then V_B would be 4.5 volts. If the value of R_2 changed slightly, so that the value of V_B decreased by a few millivolts, it would be hard to accurately make the measurement because 1 millivolt out of 4.5 volts is five parts in 10,000. However, if we add a second divider, as in Figure 1.2, then we would be measuring the difference between two similar voltages. In addition, the measurement would be less sensitive to changes in V_{BT1} since the resultant output voltage change would affect both dividers. If we now set $R_1 = R_2 = R_3 = R_4$ then $V_B = V_D$ and meter M1 would

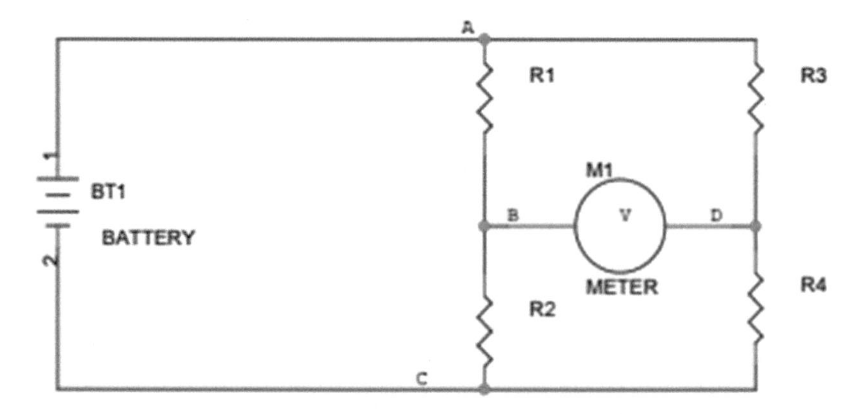

FIGURE 1.2 Wheatstone Bridge.

read 0 volts. This would be true for any battery voltage. Now if we replace R_2 with a thermistor, RT1 and replace R_4 with the appropriate potentiometer (or trimpot), for any given thermistor temperature, we could adjust potentiometer R_4 so that the bridge is balanced and reads 0 volts.

When measuring the temperature of an object, the temperature of the sensor is initially at a different value than that of the object, which in many cases would be the room temperature. Then we put the sensor in contact with the object and wait for the temperature of the sensor to change and stabilize at the new temperature before making the reading. To measure a person's body temperature the sensor is usually placed in the person's mouth. The time it takes to complete the measurement would be dependent on the thermal time constant of the sensor, which is the time it takes the sensor's temperature to reach the temperature of the body. Since it is somewhat uncomfortable for the subject to keep the sensor in their mouth, we would like to complete the reading as quickly as possible. It is possible to make the measurement quicker by extrapolating the result from a number of sampled data points over time, as the sensor is stabilizing at the new temperature to calculate the final value before the sensor temperature actually stabilizes, and report that value. When designing a modern digital thermometer, a simple microcontroller would be used that is powerful enough to quickly complete this calculation without impacting the cost of the product, thus decreasing the time it takes to complete a measurement.

S1.3 OVERVIEW OF THE EXPERIMENT

This laboratory exercise will introduce the concepts of the Wheatstone Bridge and of extrapolation. A Wheatstone Bridge will be built incorporating a thermistor, data will be sampled and the initial readings will be used to calculate the final value by extrapolation. The projected value will be compared to the actual final value.

S1.4 SAFETY NOTES

There is no health risk in this studio other than standard precautions that need to be observed in dealing with electronics.

S1.5 EQUIPMENT, TOOLS, ELECTRONIC COMPONENTS AND SOFTWARE

Additional information about the use of the items required in this studio can be found in the introduction and the appendices of this book.

Equipment

a. 9 V battery.

b. Multi-meter.

c. Data Acquisition Card (DAQ) installed into a computer.

Tools

a. Breadboard (protoboard).

b. Two BNC-to-micro clips cables.

Electronic Components

a. Thermistor EPCOS (TDK) B57164K0103J000.

b. Two 10K 1% resistors.

c. 20 kΩ trimpot (10-turn is preferred).

Software

a. MATLAB®.

S1.6 PRE-LAB QUESTIONS

Questions to be answered before starting to read and/or implement the sequence of experimental steps in the following section 1.7:

1. Do the students have knowledge of using trimpots and its terminal configuration? (e.g., the terminal that is physically in the middle is also shown in the middle in the schematic and the side ones are interchangeable).

2. Do the students know how to measure resistances using an ohmmeter and what the tolerance color codes of the resistors mean?

3. Do the students have an understanding of thermal resistance between the thermistor and the environment and how it affects the time constant of the thermistor? (Note that thermal resistance is given in units of °C/W in the datasheets of electronic components, especially for semiconductor components.)

4. Which physical properties of the temperature sensor define the time constant? (Hint: thermal resistance and thermal capacitance. Time constant is the product of the two).

5. How does the thermal capacitance relate to the mass of the sensor and the specific heat of the material it is made of?

S1.7 DETAILED EXPERIMENTAL PROCEDURE

1. Build the Wheatstone Bridge circuit shown in Figure 1.3 on a breadboard. Your completed circuit should be similar to the one in Figure 1.4.

   Take two BNC-to-micro-clips cables. Connect the red micro clip on one cable to B and on the other cable to D. Connect both black leads to C. Connect the B lead cable BNC to the first input channel of the DAQ Board (ACH0) and the D lead BNC to ACH8. The ACH0 and ACH8 are the inputs for the first differential channel of the DAQ Board. If desired another differential input pair can also be used (Figure 1.3).

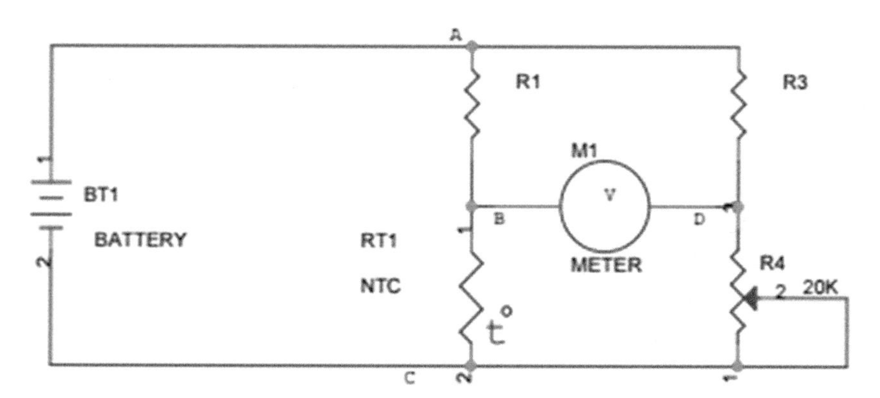

FIGURE 1.3 Wheatstone Bridge thermometer.

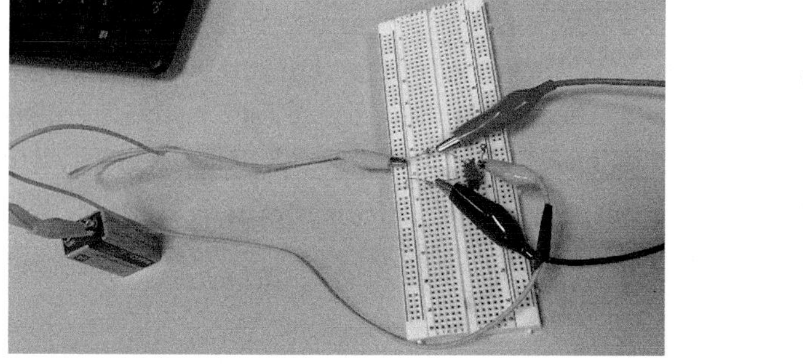

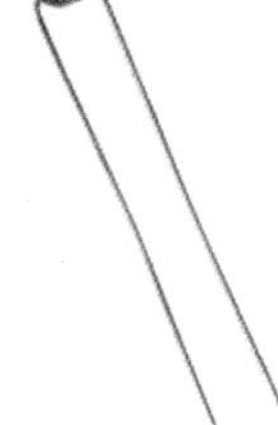

FIGURE 1.4 Typical experimental setup.

S1.7.1 Circuit Testing and Troubleshooting

- Before turning the power on, carefully check if the voltage supply is applied to the circuit with the correct polarity. Using a voltmeter with one lead connected to the ground, check the 9 V and ground voltages. It is a time saving practice to do this as the first step of troubleshooting in any circuit, analog or digital.
- Another good practice is to measure the total current that the circuit is drawing from the power supply. If the current is higher than a few mA check your wiring and component values. If the current is 1 mA or less, check for open connections.
- Using your multimeter in voltmeter setting, confirm that the potentiometer can be adjusted for 0 volts differential output between the points B and D. If it cannot be, then there is a connection issue with your circuit.

2. Open MATLAB and invoke *softscope* (or *analogInputRecorder* in newer versions of MATLAB). At the configuration menu, set the sample rate to 1000, the input type to "differential" and the input range on channel 0 to ±2.5 V.
3. Notice that the circuit needs only one 9 V power supply. While monitoring *softscope* adjust R4 for as close to zero output as possible with RT1 in free air. If you cannot accomplish this, follow the troubleshooting steps, before continuing with the experimental procedure.
4. After troubleshooting the circuit, pinch the thermistor between your thumb and forefinger. Monitor *softscope* until the waveform stabilizes. Note the time and the voltage, which should be around 1.1–1.2 volts. Release the thermistor (Figure 1.5).
5. Set *softscope* so the total time displayed is about twice the time the waveform stabilized in. After the thermistor voltage restabilizes to 0 volts, start a single acquisition and repinch the thermistor.
6. MATLAB curve fitting tools can be used to fit a curve to the data and extrapolate the end point.
7. The infinity value of the function that is used in the curve fitting tool gives the projected temperature value (Figure 1.6).

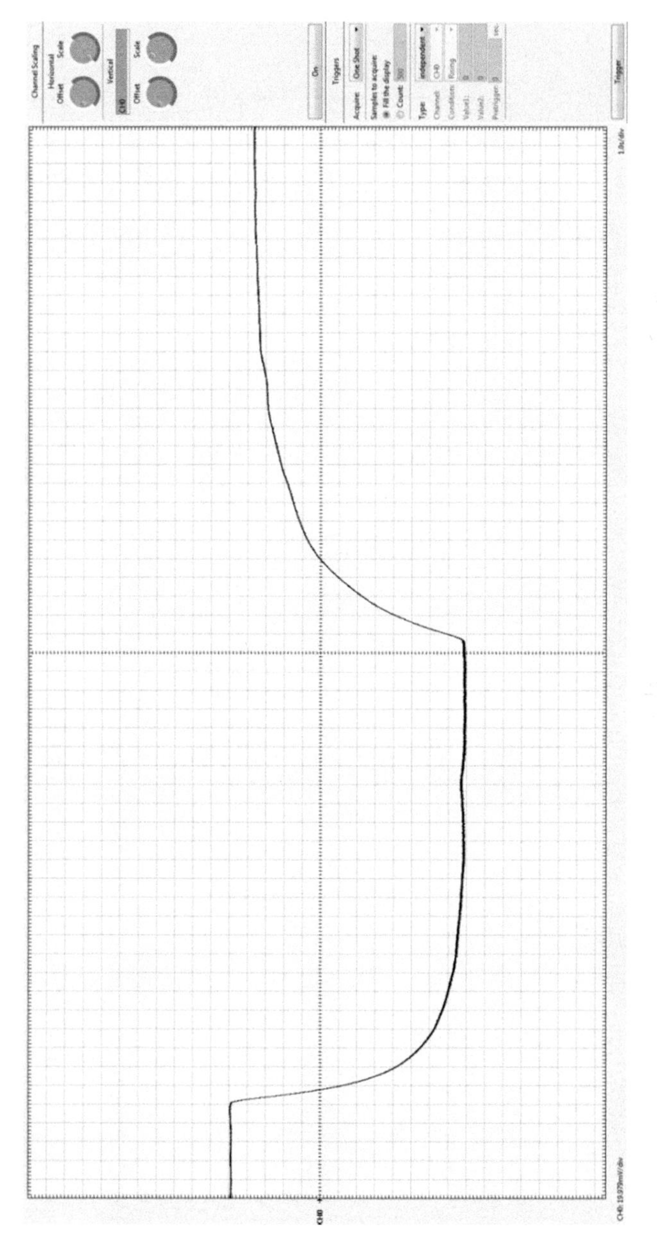

FIGURE 1.5 Typical results of pinching and releasing the thermistor. (Note: the polarity may be reversed if connections to CH0 and CH8 are reversed.)

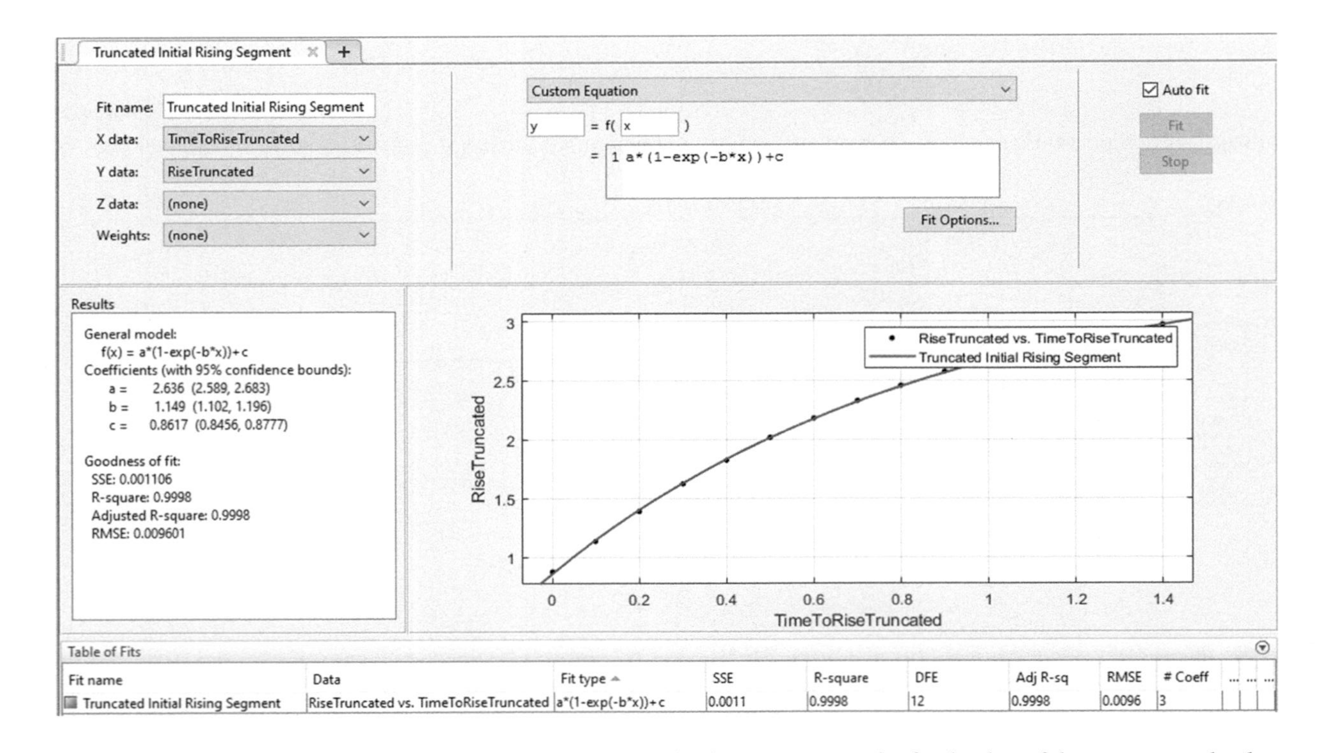

FIGURE 1.6 Estimation of the final temperature, i.e., the voltage, value by projection. The final value of the exponential voltage curve is predicted by extrapolating from the initial 63% of the step response. The exponential function fit to the data had an R^2 value of 0.9998, showing a great fit. The c and a coefficients are added to predict the steady-state voltage value of 0.8617 + 2.636 = 3.498 V. Compared to the actual steady-state value of 4.16 V, this prediction had an error of 16%.

S1.8 DATA ANALYSIS AND REPORTING

The following items are a shortlist of topics of relevance that should be included in the lab report, as a minimum for discussing the results of this studio:

1. Explain the characteristics of the Wheatstone Bridge.
2. Explain the operation of the thermistor.
3. Discuss the math behind curve fitting and function extrapolation.

S1.9 POST-LAB QUESTIONS

The following questions can be answered after carrying out the experimental steps above in section 1.7 as self-assessment of your understanding of the concepts involved in the studio or as an addition to the lab report.

1. Was the time constant the same on the rising (finger pinching) and falling (release) parts of the temperature curve? Which one is longer and why? (Hint: thermal resistance at the surface of the thermistor is not the same when the sensor is in contact with finger or air.)
2. Would the time constants be the same for different rooms (ambient) temperatures?
3. How well does a single exponential curve fit the temperature data?
4. Can the final value of the temperature be predicted from the initial section of the curve by extrapolation with curve fitting? What are the factors that can affect accuracy?

REFERENCES AND MATERIAL FOR FURTHER READING

References marked with an asterisk (*) are recommended to those interested in expanding on the content of this chapter.

1. *Curve fitting with MATLAB*:: http://www.swarthmore.edu/NatSci/echeevel/Ref/MatlabCurveFit/MatlabCftool.html
2. *NTC Datasheet*, TDK Series/Type B57164K.

STUDIO 2

Electrophysiological Amplifier

Recording Electrocardiograms Through A Breadboard

S2.1 LEARNING OBJECTIVES

The objectives of this studio are to:

1. Acquire the fundamentals of *Instrumentation Amplifiers* (IAs).
2. Recognize the superior characteristics of IAs over single-Op-Amp differential amplifiers.
3. Learn how to determine experimentally the Common-Mode-Rejection Ratio (CMRR).
4. Become familiar with the procedures to connect an IA to a subject for electrophysiological signal recording.
5. Get acquainted and implement additional circuits to transform an IA into an electrophysiological amplifier.
6. Record an electrocardiogram (ECG or EKG) from a human subject with a breadboard-based electrophysiological amplifier.

S2.2 BACKGROUND

The effects of electric fields on the human body have been known since the mid-eighteenth century, as proven by the publication of a book on the therapeutic uses of electric fields by the Italian physician Giuseppe Veratti in 1748 [1]. However, understanding of the physiological mechanisms underlying those bioelectrical effects was missing [2]. Even the very relationship between nerves and muscle contraction were the subject of heated debate until the Italian surgeon and scientist Luigi Galvani published his now famous experiment in which a distant spark caused a frog limb to contract [3]. This experiment by Galvani – the so-called *Galvani's first experiment* – is considered the start of the field of electrophysiology as we know it today [4].

Electrophysiology focuses on the study and recording of the electrical phenomena that occur in the human body. Recording of biopotentials is routinely performed in current clinical practice for electrocardiograms (ECG or EKG), electroneurograms (ENG), electromyograms (EMG) and electroretinograms (ERG). Bioelectric signals are produced by excitable cells, which are capable of generating an electric pulse – the *action potential*, Figure 2.1 – upon receiving an appropriate stimulus. Excitable cells include neural and muscular cells as well as some glandular and ciliated cells.

Mammalian cells are confined within a thin (7–15 nm) membrane made of lipids and proteins. The cellular membrane is semipermeable as it allows only for some molecules to cross into and out of the cellular cytoplasm. In particular, the membrane of excitable cells presents a number of proteins that regulate the flux of ions across the membrane. Among those transmembrane proteins, the most relevant ones to the production of action potentials are voltage-gated Na^+ and K^+ channels, as well as Na^+/K^+ ion pumps [5]. Voltage-gated channels open upon receiving a certain voltage pulse and then allow for the passive diffusion of specific ions along the channel. Ion pumps transport active ions from one side of the membrane to the other; because ion pumps are active ion transporters, adenosine triphosphate (ATP) is required for the pump to perform its ion transporting function.

Prior to receiving a stimulus, membranes of excitable cells are in what is known as their *resting* state (Figure 2.1). Cell membranes at rest maintain a steady voltage difference between their intracellular and extracellular sides; in humans the intracellular voltage is about 70 mV below the extracellular voltage, yielding a transmembrane voltage (v_M) nearing –70 mV. The Na^+/K^+ ion pumps are responsible for keeping the resting voltage by

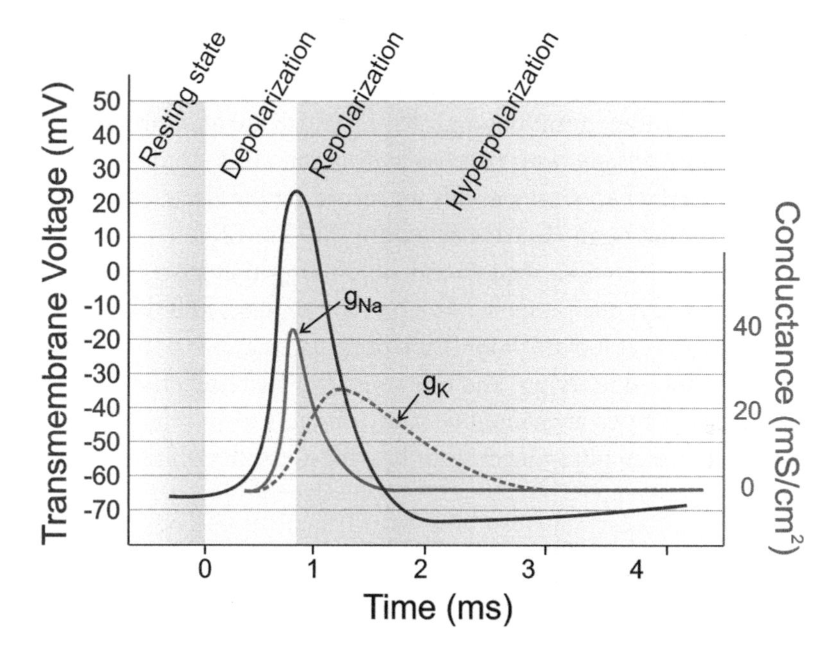

FIGURE 2.1 Schematic of a typical action potential indicating (blue) the variation of transmembrane potential and (purple) the variation of the two major membrane conductances with time: conductance of sodium channels (continuous line) and of potassium channels (dashed line). The different states in the sequence of an action potential are shown: resting state, depolarization, repolarization and hyperpolarization.

extracting three sodium ions ($3Na^+$) and introducing two potassium ions ($2K^+$) into the cytoplasm in each pumping cycle. As a result, each pumping cycle yields a net charge gain in the cytoplasm of −1 electron, in agreement with the negative sign of the recorded resting v_M.

After adequate stimulation, the voltage-gated Na^+ channels open (as indicated by the sudden increase of the Na^+ conductance, g_{Na}, in Figure 2.1) thus allowing extracellular Na^+ ions to rush into the cytoplasm, and depolarizing the cell membrane by inverting the sign of the transmembrane voltage. Voltage-gated K^+ channels also open after receiving the stimulus, letting the potassium ions diffuse out of the cell. The potassium channels, however, have a slower response than the sodium channels, as shown by comparing g_K and g_{Na} in Figure 2.1. Repolarization of the cell membrane starts when the Na^+ channels begin to close and lasts until the transmembrane voltage equals the initial resting voltage. After an action potential, the intracellular voltage reaches values lower than those at rest, taking the cell membrane to a hyperpolarized state [6].

Different excitable cells in the human body and, by extension, different excitable tissues and organs produce distinct action potentials with characteristic amplitudes, durations and shapes, which are oftentimes referred to as the *electrical signatures* of a tissue or organ. This studio focuses on implementing the necessary circuits for recording the electrical signature of the heart – that is, an electrocardiogram (ECG or EKG) [7].

The heart can be described as a double pump that distributes blood to the entire body; one pump sends blood to the lungs and the other one feeds blood to the rest of the body. Each of the two cardiac pumps consists of two chambers: one atrium and one ventricle. Therefore, the heart, as a whole, consists of two atria and two ventricles, as shown schematically in Figure 2.2. For each pumping cycle or heartbeat, the heart fills with blood

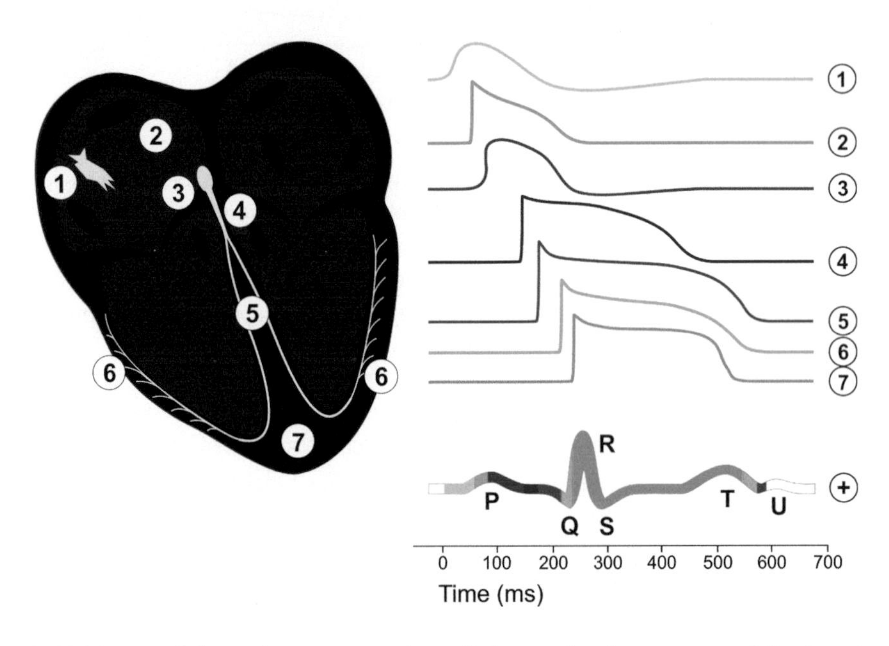

FIGURE 2.2 Schematic representation of the conduction system of the heart correlating the origin of each electrical signal in the heart with its triggering time – relative to that of other signals – and characteristic voltage signature. The trace at the bottom is the ECG waveform that results from the combination of all the cardiac signals produced by each one of the excitable cells in the heart; typical amplitude of the ECG signal is 1–3 mV. The numbers in the schematic stand for sinoatrial (SA) node (1), atrial muscle (2), atrioventricular (AV) node (3), common bundle (4), left and right bundle branches (5), Purkinje fibers (6) and ventricular muscle (7). Note: the heart is shown as one would see it in another person – i.e., the right side of a subject's heart appears on the left to the viewer and vice versa.

(*diastole*) through the atria and subsequently pushes the blood out of the ventricles (*systole*). The sequence of steps involved in a heartbeat is carefully orchestrated by the *cardiac conduction system*, which consists of the sinoatrial node (SA), the atrioventricular node (AV), the common bundle, the bundle branches, and the Purkinje fibers. The SA node is known as the physiological pacemaker of the heart because it is capable of *self-firing* – that is, it is capable of generating an action potential without receiving an external stimulus. Overall, the SA node-firing frequency is regulated by the central nervous system (CNS), which adapts the heart rate to various physiological factors such as the breathing rhythm [8].

The contraction of the heart starts with the firing of the SA node, which produces an action potential that spreads across the atria (also known as auricles) and into the AV node. The atria contract upon receiving the trigger signal from the SA node and push the blood into the ventricles through the cardiac valves. Then the AV node fires a pulse into the common bundle, the bundle branches, and ultimately, the Purkinje fibers that cause the contraction of the ventricles and the delivery of the blood outside of the heart into the bodily organs [9]. Cells in each one of the components of the cardiac conduction system fire at a different time and produce an action potential with a distinctive shape or signature (Figure 2.2). The waveform that we are used to seeing in ECGs is the combination of the action potentials produced by each one of the cells in the heart. The cardiac conduction system, despite its key role in the function of the heart, contributes only a small portion of the volume and the cells of the entire heart. By comparison, the muscle cells in the atria and the ventricles are much more numerous than the cells in the conduction system and as a result, the main factor for the recorded ECG signal is the firing of the muscle cells in atria and ventricles. The P wave in the ECG corresponds to the depolarization of the atria. The QRS complex, a component of the ECG waveform, arises from the depolarization of the ventricles; the repolarization of the atria also occurs during the time of the QRS complex, but the atrial signal is much weaker than that produced by the ventricles. The T wave is the result of the repolarization of the ventricles. Finally, the U waves are related to the repolarization of the Purkinje fibers although these waves are frequently not visible in ECGs. The ECG feature resulting from combining waves P through U is commonly referred to as *electrical systole* whereas the rest of the ECG between one U wave and the following P wave is known as *electrical diastole*.

As the depolarization process advances from the P to the U waves, the electrical signal also progresses from the atria to the ventricles. Such directionality of the electrical currents in the heart during a heartbeat can be schematically described as an arrow crossing the heart diagonally from top right to bottom left, following what is known as the *mean electrical axis* of the heart (Figure 2.3). The length of the arrow displays the *mean potential (voltage)* of the heart, which results from combining all the voltage vectors produced by all the excitable cells in the heart. The electrical signals in the heart change in time during the cardiac cycle; the mean electrical axis and mean potential of the heart are typically calculated using data from the QRS complex, when the electrical activity of the heart is maximal.

Among the multiple methods to take ECGs, the simplest ones are the *bipolar* configurations or leads, which require only two electrodes for

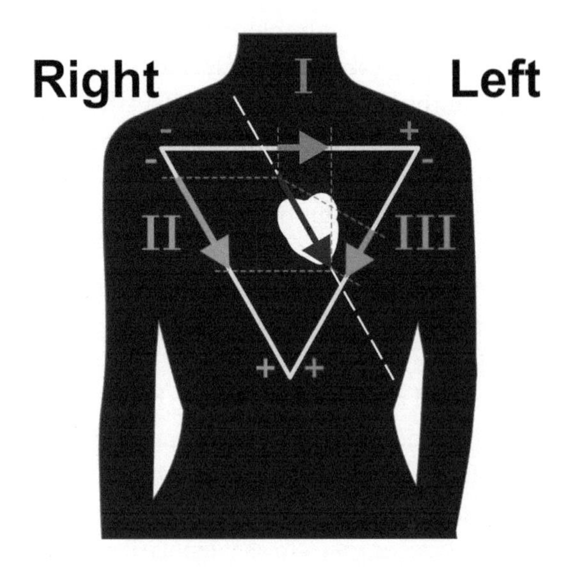

FIGURE 2.3 Schematic of the Einthoven's triangle (continuous yellow line) overlapping the heart (white in center) and the mean electrical axis of the heart (white dashed line). The three pairs of positive (+) and negative (–) electrodes of the standard bipolar leads are indicated in different colors: Lead I in orange, Lead II in blue, and Lead III in green. The reference electrodes are omitted from the schematic for simplicity. The amplitude of an R wavelength is displayed as a red arrow with the head and the tail of the arrow pointing at the distribution of positive and negative charges, respectively. The projection of the R wave onto Lead II (blue arrow) is maximal because Lead II runs in approximately the same direction as the mean electrical axis. The projections of the R wave onto Leads I (orange arrow) and III (green arrow) are shorter.

measuring the cardiac signal and one to connect the subject with the reference voltage. The most common bipolar leads are those on the front of the human body: Leads I, II, and III were established by Willem Einthoven, a Java-born Dutch physician who received a Nobel Prize in 1924 for his studies advancing electrocardiography [7]. Einthoven's triangle describes the positioning of each signal electrode for Leads I, II and III relative to the mean cardiac axis of the heart (Figure 2.3); the reference electrode is omitted from the schematic for simplicity. In Einthoven's triangle the lines connecting the electrodes are called *lead axes* and determine the direction of the measurements. Lead II runs in approximately the same direction as the mean electrical axis of the heart and as a result Lead II is able to record the QRS complex with fewer losses than Leads I and III, which are oriented at an angle to the maximum cardiac current flow (Figure 2.3).

The ECG in this studio will be recorded using Lead I, a configuration that Figure 2.3 shows with the positive electrode on the upper left side of the subject's chest, the negative electrode on the upper right side of the chest, and the ground electrode in the abdominal region. In clinical settings, for convenience to the subject, electrodes are typically positioned on wrists and legs instead of chest and abdomen (Figure 2.4). Recordings on the limbs are considered equivalent to, yet somewhat weaker than, those on the torso thanks to the conductivity of body fluids that facilitate the transport of signals from the torso to limbs.

Electrocardiography is a primary instrument for diagnosis in clinical settings because (i) it reflects multiple factors of relevance to the health and disease states of a subject, and (ii) it can be collected by means of non-invasive recording methods. Typically, ECGs are recorded from a number of electrodes adhered to the subject's skin. The presence of water-based fluids in the body provides cardiac waveforms with low-conductance pathways for traveling from the heart to the skin. *Body-surface voltages* are usually weak, with amplitudes in the mV to μV ranges, and need to be amplified before further use. The amplitudes of ECG signals, for example, are in the range of only a few mV, typically requiring amplifications in the hundreds or thousands. The type of amplifiers required for ECG signals are called *differential amplifiers* because they do not just amplify one signal or another, but they amplify the difference between two signals. In the case of the ECG, they amplify the difference between the signals collected by the positive and the negative electrodes in one bipolar lead. The simplest differential amplifier is known as *one-Op-Amp amplifier*

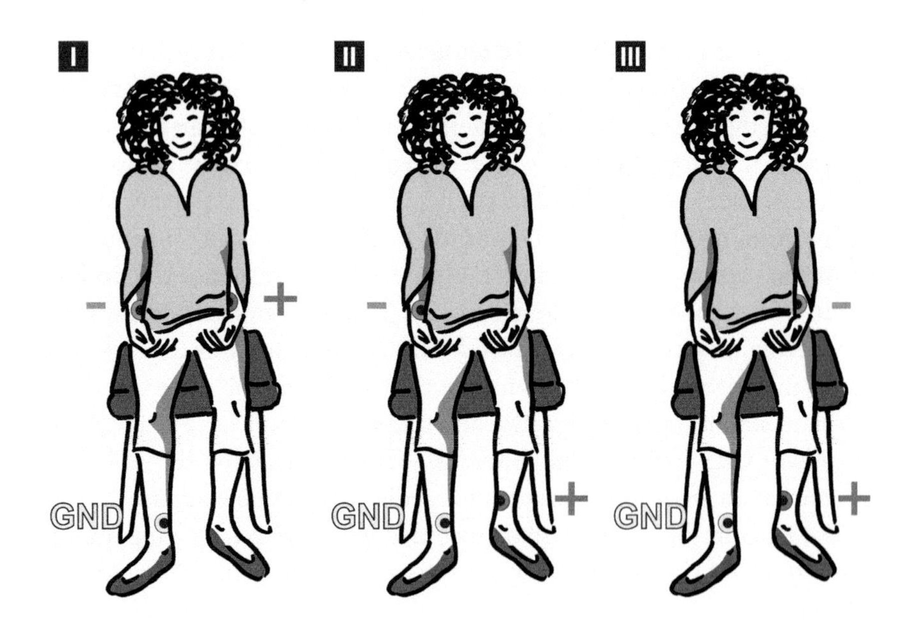

FIGURE 2.4 Schematic of the most common bipolar leads (I, II and III) as implemented typically in clinical settings – that is, with the electrodes positioned on the wrists and the legs of the subject. For easy localization, in the schematic the positive electrodes are labeled in red, the negative ones in blue, and the reference or ground ones (GND) in yellow.

(Figure 2.5B). The *transfer function* or relationship between the input signals (v_3 and v_4) and the output signal (v_o) of this amplifier is:

$$v_o = \frac{R_6}{R_3}\left(\frac{R_3 + R_4}{R_5 + R_6}\right)v_4 - \frac{R_4}{R_3}v_3 \tag{2.1}$$

If the resistors are chosen such that $R_3 = R_5$ and $R_4 = R_6$, Equation 2.1 becomes

$$v_o = \frac{R_4}{R_3}\left(v_4 - v_3\right) \tag{2.2}$$

which yields a transfer function that depends only on the values of the resistors and the difference between the two inputs. The *differential gain* of the one-Op-Amp amplifier is $G_D = R_4/R_3$. As described, this amplifier presents the advantages of (i) a differential gain easily adjusted through the values of the resistors and (ii) high rejection to common mode signals – that

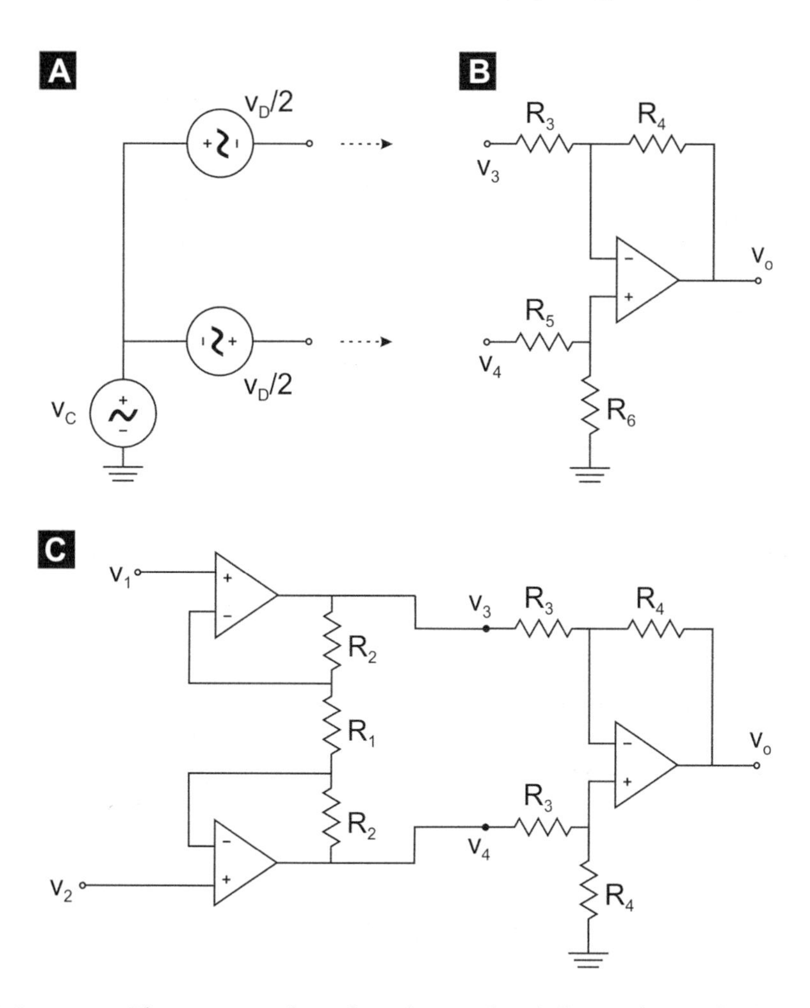

FIGURE 2.5 The circuit on the right is the simplest differential amplifier known as One-Op-Amp differential amplifier. When connected to the circuit to the left, the one-op-amp amplifier becomes an instrumentation amplifier with outstanding characteristics for bioinstrumentation.

is, the amplifier will be highly capable of rejecting signals that enter the circuit through both of the inputs simultaneously. A relevant example of common mode signal that needs to be eliminated frequently from circuits is the interference from the power line that appears in circuits as a peak at 50 or 60 Hz (depending on the country). To analyze the effect of common (v_C) and differential (v_D) mode signals in the amplifier, one applies those signals as shown in Figure 2.5A and determines how they affect the output voltage.

$$v_o = \frac{R_3R_6 - R_4R_5}{R_3(R_5 + R_6)} v_C + \frac{R_6(R_3 + R_4) + R_4(R_5 + R_6)}{2R_3(R_5 + R_6)} v_D = G_C v_C + G_D v_D \quad (2.3)$$

where G_C stands for the *common-mode gain*. The gains are strictly defined as:

$$G_C = \frac{v_o|_{v_D=0}}{v_C} \quad (2.4)$$

$$G_D = \frac{v_o|_{v_C=0}}{v_D} \quad (2.5)$$

The ability of an amplifier to reject common signals is typically quantified through the Common Mode Rejection Ratio or CMRR [10]:

$$CMRR = \frac{G_D}{G_C} \quad (2.6)$$

which can also be calculated in decibels (or dB units, section 1.2:

$$CMRR = 20\log_{10}\left(\frac{G_D}{G_C}\right) \quad (2.7)$$

In the case of the one-Op-Amp amplifier,

$$CMRR = \frac{R_6(R_3 + R_4) + R_4(R_5 + R_6)}{2(R_3R_6 - R_4R_5)} \quad (2.8)$$

The CMRR is maximum (infinity) when $R_3R_6 = R_4R_5$, which includes the case chosen earlier $R_3 = R_5$ and $R_4 = R_6$ (Equation 2.2). In reality, however, the resistor pairs will never be exactly identical and the CMRR will be a finite number.

A major limitation of the one-Op-Amp amplifier is its low input impedance. Despite the large input impedance of the operational amplifier (Z_{OA}), input signals encounter paths of low resistance through R_3 and R_4, which are much smaller than Z_{OA} and the equivalent impedance of the electrodes, which typically is about 10 kΩ for gel-based, floating-metal, disposable electrodes in the ECG frequency range [11]; notably, the impedance of gel-based electrodes increases dramatically as they dry. An alternative configuration is the *three-Op-Amp differential amplifier*, also known as *instrumentation amplifier* (or INA; Figure 2.5C). Different to the previous one, this amplifier presents input signals with a large impedance – i.e., that of the operational amplifiers. Also, the differential gain is improved as shown by the transfer function:

$$v_o = \frac{R_1 + 2R_2}{R_1}\frac{R_4}{R_3}\left(v_1 - v_2\right) \quad (2.9)$$

Note that the instrumentation amplifier includes the previous one-Op-Amp amplifier and thus it is not surprising that the differential gains of both are related through a proportionality factor of $(R_1 + 2R_2)/R_1$. The CMRR of the INA also increases by that factor when compared to the one-Op-Amplifier:

$$CMRR = \left(\frac{2R_2 + R_1}{R_1}\right)\frac{R_6\left(R_3 + R_4\right) + R_4\left(R_5 + R_6\right)}{2\left(R_3R_6 - R_4R_5\right)} \quad (2.10)$$

This studio will use an instrumentation amplifier with $R_3 = R_5$ and $R_4 = R_6$ for collecting an ECG (Figure 2.6). Additional to the amplifying stage, the circuit in this studio includes a series of filters that will minimize the noise accompanying the collected ECG signal. Filters are able to separate certain parts of a signal, typically according to their frequency, and process them differently – for example, by blocking one part while leaving the rest intact. The frequencies of interest in an ECG are between 0.05 and 100 Hz. Importantly, other bioelectric signals have different frequencies of interest: electroencephalograms (EEG) and electromyograms (EMG), for example, use frequency ranges of 3–32 Hz and 10–500 Hz, respectively.

One *high-pass filter* is located at each of the inputs of the circuit (Figure 2.6). As the name indicates, high-pass filters allow for high frequencies to pass whereas the low frequencies are attenuated. The frequency that separates attenuated from non-attenuated frequencies is the *corner* or *cutoff frequency* (f_C). The high-pass filters used in this studio, known as *first-order passive high-pass filters*, have a cutoff frequency of

$$f_C = \frac{1}{2\pi RC} \quad (2.11)$$

where R and C are the values of the resistors and capacitor that form part of the filter. At low frequencies the impedance of the capacitor increases, thus hindering the pass of low frequency signals; conversely, at high

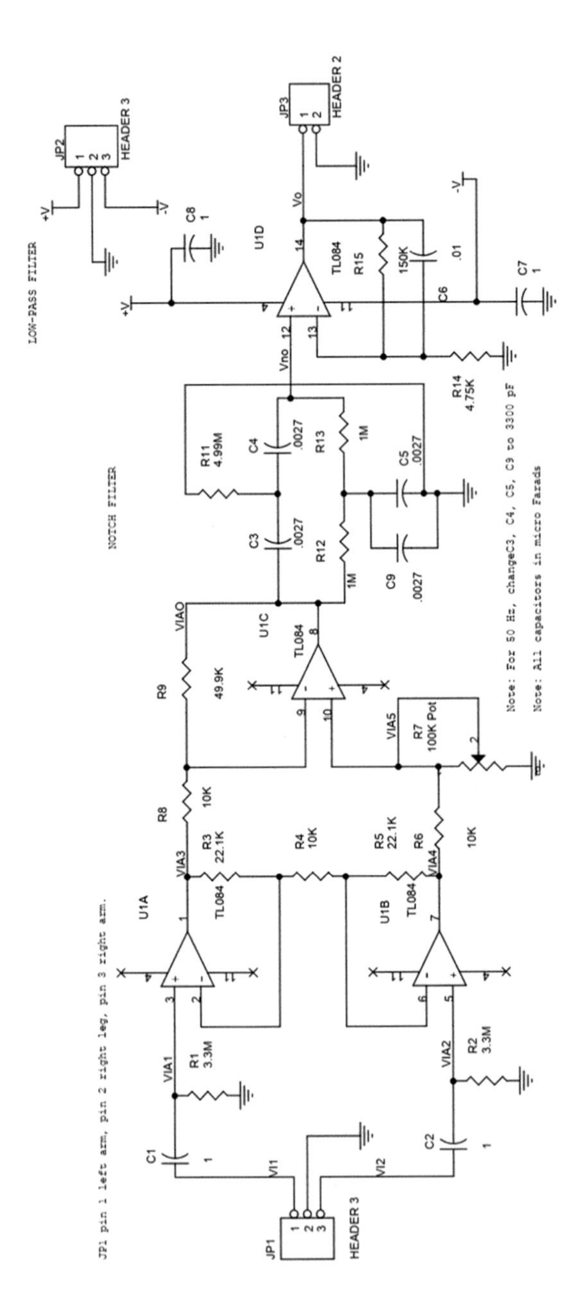

FIGURE 2.6 Schematic of the circuit for the electrocardiograph in Studio 2. The block diagram indicates the different stages of signal processing in the circuit. The output of the circuit will be collected and displayed by an oscilloscope.

frequencies, the capacitor behaves similarly to a short-circuit thus favoring the pass of high-frequency signals. High-pass filters are crucial in situations where a small AC signal appears added to a large DC signal (zero frequency) such as the one associated to the half-cell potential of the electrodes – as discussed at length in Studio 13. High-pass filters are able to block the DC signal while allowing the AC signal with frequency higher than f_C to advance towards the amplification stage. For this studio, the cutoff frequency of the high-pass filters comes determined by the lowest frequency of interest of the ECG signal.

A *low-pass filter* is located at the output of the circuit, which blocks frequencies higher than the cutoff frequency. This type of low-pass filter (*first-order passive low-pass filter*) is identical to the high-pass filters above except that the positions of the capacitor and the resistor are exchanged. The cutoff frequency of these low-pass filters is also given by Equation 2.11 and comes determined by the highest frequency of interest of an ECG signal. It is important to notice that the combination of the input high-pass filters and the output low-pass filter will be able to attenuate all frequencies outside the range of interest.

The third and last type of filter used in this studio is a *notch filter*, which strongly attenuates a narrow band of frequencies. This filter aims at blocking any power line interference signal (at 50 or 60 Hz) that remains after the common-mode rejection by the instrumentation amplifier. The configuration chosen here is called the twin-T notch filter, as its circuital schematic resembles two Ts mirroring each other. One T behaves as a low-pass filter whereas the other T behaves as a high-pass filter. These two filters have cutoff frequencies that are very close but do not overlap; as a result, the range of frequencies located between the two cutoff frequencies will be attenuated by the two filters simultaneously, creating the characteristic notch. The expression for calculating the notch frequency of the twin-T filter is given by Equation 2.11. From an experimental point of view, it is important to note that the efficacy of the twin-T filter is limited by the tolerance of its components—that is, the depth of the notch is reduced dramatically by mismatching values of the resistors and capacitors in the filter.

S2.3 OVERVIEW OF THE EXPERIMENT

This studio will implement the first Instrumentation Amplifier (IA) in this handbook. Most importantly, this studio will discuss in depth the issues that arise when an IA is connected to a human subject and how to address them. The IA will be built on a breadboard together with three types of signal filters (low-pass, high-pass, and a 60 Hz notch filter) that will eliminate the noise blurring the electrophysiological signal of interest – in this chapter an EKG – collected by the electrodes on the human subject (Figure 2.1). Extensive tips on circuit troubleshooting will be provided to support this first IA-based studio in the book.

S2.4 SAFETY NOTES

This studio involves the recording of physiological signals from human subjects. Depending on regulations at your institution, you may need to acquire an official approval from the Institutional Review Board (IRB) before conducting these experiments.

It is extremely important that the subject is not connected directly to the ground of any of the instruments used during this studio – e.g., oscilloscope, voltage supply. In hospitals the clinical equipment has isolated grounds that prevent the passage of AC or DC current from such equipment to the patients. Isolated grounds are a safety measure to protect patients in case of faulty equipment. A direct connection between the subject and the ground of a faulty apparatus may result in fatality. Additionally, all electric equipment in the laboratory should be tested for safety prior to use; the host institution is typically responsible for such safety measures.

S2.5 EQUIPMENT, TOOLS, ELECTRONIC COMPONENTS AND SOFTWARE

Additional information about the use of the items required in this studio can be found in Studio 1 and in the Appendices.

Equipment

- Breadboard.
- Double voltage supply (±15V).
- Multimeter.
- Oscilloscope.
- Sinusoidal signal generator.

Tools

- BNC-to-microclip cables.
- BNC-to-banana cables.

- Banana-to-banana cables.
- Jumper wires for the breadboard.

Electronic Components (remember to download the available datasheets)

a. 1 Quad-Operational Amplifier, TL084CP.
b. Various 1/4W resistors and capacitors (low leakage preferred).

   2 – 3.3M.

   1 – 150K.

   2 – 22.1K 1%.

   2 – 10K 1%.

   1 – 49.9K 1%.

   2 – 1 M 1%.

   1 – 4.7 K.

   4 – 1 μF 5%.

   4 – .0027 μF 5% (60 Hz).

   4 – .0033 μF 5% (50 Hz).

   1 – .01 μF.

   1 – 100 K potentiometer.
c. 1 – 2 pin header (optional).

   2 – 3 pin header (optional).
d. Disposable ECG electrodes.
   Note: It is important to use 1% resistors where specified, and 5% capacitors, otherwise circuit performance will be compromised.

S2.6 PRE-LAB QUESTIONS

Questions to be answered before starting reading and/or implementing the sequence of experimental steps in the following section 2.7.

1. For a first-order high-pass filter like the ones in Figure 2.3, calculate the cutoff frequency (fc, Equation 2.11) given by a 3.3 MΩ resistor and a 1 μF capacitor.

2. For a first-order low-pass filter like the one in Figure 2.3, calculate the cutoff frequency (fc, Equation 2.11) given by a 1.5 kΩ resistor and a 1 μF capacitor.

3. Which range of frequencies will be allowed to pass by both the high- and the low-pass filters described in the previous two questions? Discuss whether such range is fit for recording an ECG.

S2.7 DETAILED EXPERIMENTAL PROCEDURE

1. Measure with the multimeter the actual values of all the resistors and capacitors needed in this studio – see section 2.5 – and record those measurements for later use.

   TIP: The introduction provides a summary of best practices for using the multimeter.

   ATTENTION: Mistakes often arise from difficulties deciphering the color of resistor stripes or the letters and numbers in capacitor labels. The best practice is to measure the value of each electronic component before placing it on the breadboard.

2. Download and explore the datasheet of the chosen Op-Amp – e.g., TL084 or LM324.

3. Build the Instrumentation amplifier (IA) – i.e., INSTRUMENTATION AMPLIFIER block in Figure 2.6. Use a supply voltage of ±15 V for the Op-Amp.

   TIP: The introduction provides a summary of best practices for using the breadboard.

   TIP: Slightly higher or lower values can be used for the resistors as long as the resistors with identical values remain equal to each other.

4. Calculate the theoretical gain of the IA according to the values of the resistors used.

5. Determine experimentally the gain of the instrumentation amplifier.

   a. Using the signal generator, apply a small 1 kHz sinusoidal signal – e.g., 100 mVpp, where 'pp' indicates peak-to-peak amplitude – to v_{IA1} and connect v_{IA2} to ground.

ATTENTION: Before turning the power on, ensure the values of the resistors and capacitors are correct.

ATTENTION: Before turning the power on, ensure the location of the resistors and capacitors are correct. A convenient method to certify the components are in their right location is, (i) to verify each connection on the breadboard and then, (ii) mark it on a printout of the schematic one connection at a time.

ATTENTION: Before turning the power on, confirm that the supply voltage applied to the operational amplifier (Op-Amp) has the correct polarity. Modern Op-Amps are able to tolerate many types of circuital errors such as short-circuited outputs, excessive voltage applied to the inputs, etc. However, when the supply voltage is applied to the Op-Amp with the reverse polarity the chip burns in a fraction of a second. Notably, if the Op-Amp burns quickly, the plastic package covering the chip may not burn or even heat up.

TIP: Section I.6 provides a summary of best practices for using the signal generator.

b. Display both the input and the output of the instrumentation amplifier on the oscilloscope by connecting v_{IA1} and v_{IAO} to channels 1 and 2, respectively, of the oscilloscope.

TIP: Section I.3 provides a summary of best practices for using the oscilloscope.

TIP: As described by Equation 2.9, the IA is not inverting and therefore its input (v_{IA1}) and output (v_{IAO}) signals will present the same phase.

ATTENTION: If a sinusoidal signal is not detected at the output of the amplifier and the connections and electronic components have been solidly verified (as suggested in S2.7, step 5a), it is possible that an offset is masking v_{IAO} or that the amplifier is saturated:

- The best way to discern small AC signals that are superimposed onto large DC voltages is to activate the *AC setting* of the oscilloscope, which will block the DC signal and display the AC signal only. As a result, while using the AC setting, the effects of display magnifiers in the oscilloscope, such as the vertical gain button, will magnify the AC signal only.

- If v_{IAO} shows a DC value near the positive or negative supply levels of the Op-Amps, the amplifier is most likely in saturation. Another indicator of saturation is when the positive and negative inputs of the Op-Amp (v_{IA4} and v_{IA3}) do not present exactly the same voltage: if the difference between the two inputs is >1 mV (approx.), the amplifier is definitely saturated, given that current Op-Amps present open-circuit gains are typically $>10^8$. When the amplifier is in saturation, and after verifying all connections are correct, one should consider replacing the semiconductor components of the circuit (i.e., Op-Amps, transistors, diodes, etc.) since it is easier to damage semiconductor elements than resistors or capacitors. Sometimes the breadboard holes under the semiconductor components become damaged with the component. It is therefore safe to set the new semiconductors onto new holes of the breadboard.

c. Determine the IA gain experimentally by dividing the amplitude of the output signal (v_{IA1}) by the amplitude of the input signal (v_{IA1}). Verify the experimental and theoretical values for the IA gain are similar.

6. Determine experimentally the CMRR at 1 kHz by following these steps:

a. Apply a large 1 kHz sinusoidal signal, e.g., 5Vpp, to both inputs (v_{IA1} and v_{IA2}) simultaneously – this signal would be the common-mode signal (v_C) in Figure 2.5A; in this configuration v_D would be null. First connect the inputs together and then link them to the positive terminal of the signal generator. The ground of the signal generator is connected to the ground of the circuit.

ATTENTION: Before turning the power on, confirm that the supply voltage applied to the operational amplifier (Op-Amp) has the correct polarity.

TIP: Section 1.3 provides a summary of best practices for using the signal generator and the oscilloscope, as well as information on the use of BNC-to-microclip cables.

b. Adjust the 100 kΩ potentiometer to minimize the output voltage for this common-mode input signal. Measure the input and the output peak-to-peak amplitudes with the oscilloscope and divide the output (v_{IAO}) by the input signal amplitude (v_C) to find the common-mode gain (Equation 2.4).

c. Maintain the 100 kΩ potentiometer as set in S2.7, step 6b. Apply a small 1 kHz sinusoidal signal (v_D), e.g., 100 mVpp, to one of the inputs and connect the other input to ground. Determine the differential voltage gain (G_D, Equation 2.5). This configuration resembles the one in Figure 2.5A, with $v_C = 0$.

d. Determine the CMRR (in dB units, Equation 2.7) using the measurements taken in S2.7, steps 6a–c.

7. Repeat S2.7, step 6 for 100 Hz first, and then for 10 kHz. Plot the experimental values of CMRR (vertical axis) against frequency (horizontal axis).

8. Build the high-pass filters (HIGH-PASS FILTERS block in Figure 2.6) in a separate region of the breadboard.

   TIP: Review your answers to the Pre-lab questions (section 2.6), which assessed how the suggested values of the components relate to the cutoff frequency of the filters.

   ATTENTION: Ceramic capacitors do not come in the microfarad range required here. Electrolytic capacitors do reach the microfarad range but their performance is limited by large leakage currents. Plastic film capacitors are therefore the most suitable for this circuit as they provide, (i) large capacitance values, (ii) in a small package, (iii) with low leakage currents. Additional information on the different types of capacitors available can be found in section 1.5.

9. Test the high-pass filters by applying a large sinusoidal signal, e.g., 5Vpp, to both inputs (v_{I1} and v_{I2}) simultaneously and reading the outputs (v_{IA1} and v_{IA2}) with the oscilloscope. The ground of the signal generator is connected to the ground of the filters. Apply signals

at different frequencies – e.g., at the calculated f_C, two frequencies under f_C, and two above f_C.

TIP: The two filters can be tested simultaneously by feeding each of the output signals to a different channel of the oscilloscope.

10. Plot $log(v_{IA1}/v_{I1})$ and $log(v_{IA2}/v_{I2})$ against frequency and certify that the filters successfully attenuate lower frequencies, as expected from high-pass filters.
11. Connect each filter to a different input of the IA as shown in Figure 2.6.
12. Build the circuit in Figure 2.7 in a separate area of the breadboard. Before connecting the breadboard to a subject (Figure 2.6), the ECG circuit will be tested with a circuit that emulates two adhesive bio-electrodes collecting ECG signals from the skin of a subject (Figure 2.7). The subject-emulating circuit presents two different resistors because even identical bio-electrodes may yield different impedances when adhered to the skin. The impedance of the skin is not uniform and depends greatly on humidity [12].

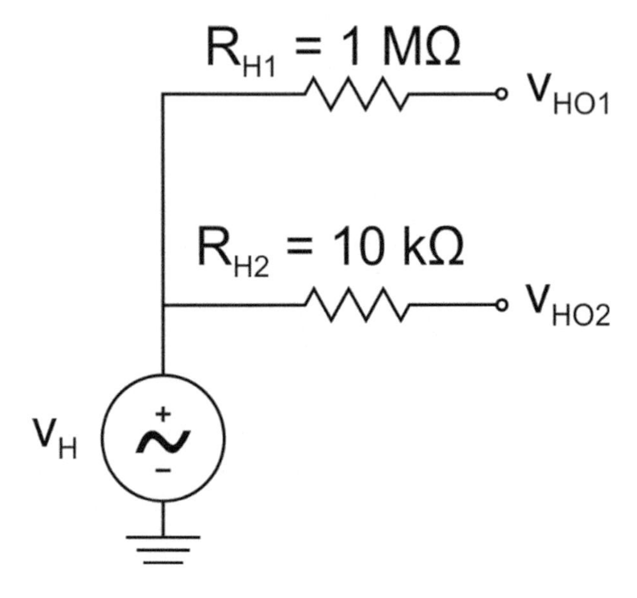

FIGURE 2.7 Schematic of a circuit that emulates two adhesive bio-electrodes adhered to the skin of a subject. The two resistors model the different impedances resulting from recording biopotentials in different areas of the skin.

13. Connect the two circuits by connecting the output signals v_{HO1} and v_{HO2} of the subject-emulating circuit to the inputs of the amplifier: v_{I1} and v_{I2}, respectively.

14. Measure the 1 kHz common-mode signal of the subject-emulating circuit connected to the ECG circuit. This step follows a similar procedure to that of S2.7, step 4a: first, apply a large (e.g., 5 Vpp) 1 kHz sinusoidal signal to R_{H1} and R_{H2} simultaneously – i.e., as v_H in Figure 2.7 – and second, measure the peak-to-peak amplitudes of both v_H and v_{IAO} using the oscilloscope. Using these measurements and assuming that the differential-mode gain remains the same as in S2.7, step6, calculate the CMRR in dB units (Equation 2.7).

    ATTENTION: For the length of this step, avoid altering the position of the rotating screw or wiper of the 100 kΩ potentiometer.

    TIP: Similar to S2.7, step 6, this step can be repeated at 100 Hz and 10 kHz in order to determine the CMRR in the whole range of frequencies of interest in ECG.

15. Build the notch filter (NOTCH FILTER block in Figure 2.6) in a separate area of the breadboard. Calculate the cutoff frequency (fc) of the notch filter according to Equation 2.11.

16. Apply a 1- Vpp sinusoidal signal to the input (v_{IAO}) of the filter. Display both the input and the output (v_{NO}) using the oscilloscope. Vary the frequency of the input signal between 50–70 Hz in steps of 1 Hz.

    ATTENTION: Depending on the amplifying gain, the output of the IA may saturate. In the event of saturation, reduce the amplitude of the input signal below 1 Vpp until the output enters the linear range.

    ATTENTION: The output voltage of the filter is expected to drop sharply to less than 10% of its maximum at around fc while staying near its maximum for the rest of the spectrum.

    TIP: If a sharp decline (notch) in the output is not observed within a few Hz around fc, the most likely explanation is that the capacitors and/or resistors are not well matched – that is, the actual values of the components are too different than those required.

Review your measurements (S2.7, step 1) and replace unmatching components by new ones that yield a better match.

17. Connect the input of the notch filter to the output of the instrumentation amplifier (v_{IAO}), as described in Figure 2.6.

18. Build the low-pass filter (LOW-PASS FILTER block in Figure 2.6).

    TIP: Review your answers to the Pre-lab questions (section 2.6), which assessed how the suggested values of the components relate to the cutoff frequency of this filter.

    ATTENTION: Ceramic capacitors do not come in the microfarad range required here. Electrolytic capacitors do reach the microfarad range but their performance is limited by large leakage currents. Plastic film capacitors are therefore the most suitable for this circuit as they provide, (i) large capacitance values, (ii) in a small package, (iii) with low leakage currents. Additional information on the different types of capacitors available can be found in Appendix I.

19. Test the low-pass filter by applying a large sinusoidal signal, e.g. 5 Vpp, to the input (v_{NO}) and reading the output (v_O) with the oscilloscope. The ground of the signal generator is connected to the ground of the filter. Apply signals at different frequencies – e.g., at the calculated f_C, two frequencies under f_C, and two above f_C.

20. Plot $log(v_O/v_{NO})$ against frequency and certify that the filter successfully attenuates higher frequencies, as expected from low-pass filters.

21. After asserting the characteristics of the low-pass filter, connect the input of the filter to the output of the notch filter (v_{NO}), as described in Figure 2.6.

22. Place three disposable electrodes on the subject – e.g., a member of your team – following the Lead II configuration (Figure 2.4): positive electrode above the inner side of the left ankle, negative electrode on the inner side of the right wrist, and ground electrode above the inner side of the right ankle.

23. Disconnect the subject-emulating circuit (Figure 2.7) and instead, connect the electrodes on the subject to v_{I1} and v_{I2}.

ATTENTION: The quality of the ECG recording depends dramatically on the contact of the electrodes with the skin of the subject, which can be optimized by following these directions:

- Have the subject relaxed, in a comfortable position – e.g., in the so-called *supine position*, the subject sits down, with legs flexed at knee and feet flat on the floor, with arms relaxed on the side of the body, and hands apart, resting on the legs with palms up.
- Clean the area of the skin where the electrodes will be placed with soap or alcohol; additionally, the skin can be scrubbed gently to remove any dead cells.
- If the electrode is dry, apply a drop of gel to it.
- Remove any jewelry located nearby the location of the electrode.
- Confirm that all the electrodes have gel in sufficient quantity and are conveniently moist – otherwise unacceptable levels of the 60 Hz interfering signal may appear at the output. Apply additional conductive gel to the electrodes when needed.
- Apply the electrodes onto the skin some time (five minutes, at least) before the start of the measurements.
- Ensure that the cables connecting the electrodes to the breadboard are supported so that they are not pulling on the electrodes.

• Check all the connections are strong, including cable clips with electrodes – weak or faulty connections frequently result in flat-lines or noisy readings.

TIP: When recording the ECG, discard the first couple of heartbeats, as the signal may take a few seconds to stabilize.

TIP: The amplitude of the 60 Hz interference signal may be reduced further by adjusting the 100 kΩ potentiometer (Figure 2.6).

24. Modify the lower corner frequency of the filter by changing the capacitor or resistor value. Move the corner frequency from 0.05 to 1 Hz. How does this affect the slow frequency components in the ECG waveform (P and T waves)?

25. Now move the high corner of 100 Hz down to 10 Hz again by replacing the capacitance with a ten times higher value. How does that affect the fast-changing QRS complex?

    ATTENTION: Component values different than those suggested could also be fitting to produce the cutoff frequency required by ECG recordings. The value of resistors R_{I1} and R_{I2}, however, needs to be high compared to the impedance of the electrodes, which is typically around tens of kΩ. As discussed earlier for the instrumentation amplifiers (Section 2.2), it is important that the input impedance is much larger than that of the connecting elements – in this case, the electrodes. Low input impedances result in degradation of the CMRR of the ECG circuit. A resistance higher than 1 MΩ is recommended at all times for R_{I1} and R_{I2}.

26. Disconnect the cables from the electrodes and carefully peel off the electrodes from the subject. Discard the disposables properly.

    TIP: Wash with soap or alcohol any gel residues that the electrodes may have left behind. Sometimes the electrodes may leave a ring-like mark on the subject's skin. Marks will spontaneously disappear in few hours without requiring any action.

    ATTENTION: Review the rules that your institution may have regarding disposal of biomedical materials such as disposable electrodes.

S2.8 DATA ANALYSIS AND REPORTING

The following items are topics of relevance that should be included in the lab report for discussing the results of this studio.

1. Explain what "common-mode gain" and "differential-mode gain" are and how they are relevant to the performance of an instrumentation amplifier.

2. State whether the common mode rejection ratio (CMRR) depends on frequency and justify your answer.

3. Rationalize the importance of having high CMRR over the frequency range of a source of noise or interference relevant to biopotential recording – e.g., the power-line signal at 60 Hz.

4. Compare your 1 kHz CMRR measurements in S2.7, steps 6 and 14, and justify any similarities or differences between them.
5. Summarize how the ECG waveform is affected by the low and high cutoff (corner) frequencies of the filters in the ECG system.
6. Outline the effects caused by electrodes drying out after prolonged use.
7. Describe any troubleshooting you performed to identify a problem and how you fixed the problem.

S2.9 POST-LAB QUESTIONS

The following questions can be answered after carrying out the experimental steps above in section 2.7, as a self-assessment of your understanding of the concepts involved in the studio or as an addition to the lab report.

1. Discuss the reasons for which disposable electrodes oftentimes present very different impedance values from one another and under different experimental conditions such as frequency, humidity, or the properties of the skin.
2. After downloading the corresponding datasheets, determine the CMRR values exhibited by commercial instrumentation amplifiers such as INA114 or AD620. Compare these commercial CMRR values to those you measured in this studio. Research how such high CMRR values are achieved with commercial amplifiers. Discuss the advantages of integrated instrumentation amplifiers over amplifiers built with discrete components such as individual resistors and operational amplifiers.
3. Note that the Instrumentation Amplifier you built using TL084 may not have a better CMRR than TL084 itself. Compare the CMRR value given in the datasheet of TL084 and the CMRR value you measured from the circuit. Why can you not achieve the catalog value of CMRR for TL084 when you build an instrumentation amplifier with discrete components? (Hint: In commercial amplifiers, the components that determine the gains through the positive and negative pathways are matched carefully in order to cancel the common-mode signal at the output.)

S2.10 ADDITIONAL EXPERIMENTAL ACTIVITIES

- Collect the ECG signal using Lead II and/or Lead III configurations. The sequence of steps for Lead I (described in section 2.7) is also useful for the Leads II and III, except for S2.7, point 22. Place the electrodes in the chosen configuration: Lead I or Lead III (Figure 2.4). Compare the ECG collected with multiple leads and discuss the differences between them.
- Compare the ECG signal at rest and after exercise. After recording the ECG of a subject, ask him/her to run upstairs and back a couple of times in order to accelerate the heart rate. (If stairs are not available, the subject can also run for a few minutes, do some push-ups or jumping-jacks, etc.) Have the subject sit down in the supine position (described in S2.7, step 21) and immediately connect the cables to the electrodes and start recording. Discuss the differences observed between the ECG at rest and the ECG after working out.

  TIP: Before start exercising, remove the cables from the subject for comfortable movement. It is important to keep the electrodes adhered so that it is easy and fast to reconnect the cables to the electrodes and restart the measurements. The acceleration of the heartbeat due to exercising is temporary. Recording by ECG needs to start as soon as possible after working out in order to capture the effects of exercising in the heartbeat of the subject.

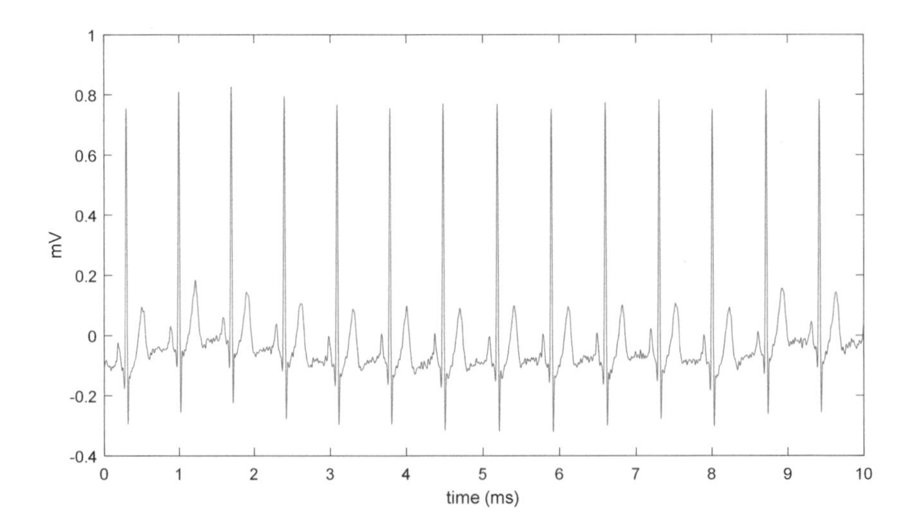

FIGURE 2.8 Example of a signal collected with the ECG in Figure 3.6, as displayed on the screen of the oscilloscope.

- Test the effect of large input impedances of the CMRR of an instrumentation amplifier. Disconnect the electrodes from the subject and instead, connect the subject-emulating circuit (Figure 2.8) to v_{I1} and v_{I2}. Disconnect the notch filter from the instrumentation amplifier in order to obtain results comparable to those collected in S2.7, step 14. Exchange resistors R_1 and R_2 by resistors with values below 1 MΩ; the resistors replacing R_1 and R_2 need to have the same resistance. Repeat S2.7, step 14 and compare this CMRR with the one recorded with $R_1 = R_2 = 3.3$ MΩ. Discuss any differences between the two measured CMRR.

REFERENCES AND MATERIALS FOR FURTHER READING

References marked with an asterisk (*) are recommended to those interested in expanding on the content of this chapter.

1. Veratti, G. *Osservazioni fisico-mediche intorno alla elettricitá*. Lelio dalla Volpe, Bologna, 1748.
2. Bressadola, M. Medicine and science in the life of Luigi Galvani (1737–1798), *Brain Res. Bull.* (1998) 46(5):367–380.
3. Galvani, L. De viribus electricitatis in motu musculari commentarius, *De Bononiensi Scientiarum et Artium Instituto atque Academia Commentarii* (1791) 7:363–418.
4. Piccolino, M. Animal electricity and the birth of electrophysiology: the legacy of Luigi Galvani, *Brain Res. Bull.* (1998) 46(5):381–407.
5. Gadsby, D.C. Ion channels vs. ion pumps: the principal difference, in principle, *Nat. Rev. Mol. Cell Biol.* (2009) 10:344–352.
6. Clark, Jr, J.W. The origin of biopotentials, Chapter 4 in *Medical Instrumentation*, ed. J.G. Webster, 4th ed. John Wiley and Sons, New York, 2010.
7. Rivera-Ruiz, M.; Cajavilca, C.; Varon, J. Einthoven's string galvanometer: the first electrocardiograph, *Texas Heart Inst. J.* (2008) 35(2):174–178.
8. Iaizzo, P.A.; Fitzgerald, K. Autonomic nervous system, Chapter 14 in *Handbook of Cardiac Anatomy, Physiology, and Devices*, ed. P.A. Iaizzo, 3rd ed. Springer, New York, 2015.
9. *Marin, G.; Cucchietti, F.M.; Vazquez, M.; Tripiana, C. 2012 International Science & Engineering Visualization Challenge—Video First Place Winner: Alya Red. A computational heart, *Science* (2013) 339(6119):518–519. Link to free-access video: http://video.sciencemag.org/Featured/2127025911001/1
10. Pallàs-Areny, R.; Webster, J.G. *Sensors and Signal Conditioning*, 2nd ed. John Wiley and Sons, New York, 2010.
11. Chi, Y.M.; Jung, T.-P.; Cauwenberghs, G. Dry-contact and noncontact biopotential electrodes: methodological review, *IEEE Rev. Biomed. Eng.* (2010) 3:106–119.
12. *Neumann, M.R. The electrode-skin interface and motion artifact, Section 5.5 in *Medical Instrumentation*, ed. J.G. Webster, 4th ed. John Wiley and Sons, New York, 2010.

Studio 3

Small Signal Rectifier-Averager for EMG Signals

S3.1 BACKGROUND

EMG analysis can be aided by obtaining the absolute value of the signal being analyzed and low-pass filtering the resultant signal to obtain the envelope of the signal. This is sometimes referred to as "rectifier-averager" since averaging the signal in a moving time window has a similar effect as the low-pass filter. This process, done before digitization, allows for a lower sampling rate than if implemented post sampling, because the envelope signal is limited to much lower frequencies than the raw signal. The resultant signal is useful in observing how the instantaneous signal power changes as a function of time. For instance, the rectified-averaged EMG signals correlate well with the force generated by the muscle.

A single rectifier diode could be used to take the positive half of the signals without any active electronics, i.e., Op-Amps. However, a semiconductor diode requires a minimum of ~0.6 V for silicon and ~0.2 V for germanium diodes across their terminals to start conducting in a forward direction. Any signal component below the diode opening voltage would be blocked by the rectifier and distort the signal output at low levels of the input. This would be true even if the EMG signals are amplified first before being sent through the rectifier diode. The circuit to be built in this studio, and the others like this, avoids this problem by including an Op-Amp in the design. Notice that these circuits would rectify the entire signal down to the microvolt level, hence the prefix "small-signal" in the title.

S3.2 OVERVIEW OF THE EXPERIMENT

This studio will implement a small signal rectifier (absolute value) circuit followed by a low-pass filter. After testing the filter and the rectifier with a generic sinusoidal signal, the circuit operation will be verified using an actual EMG signal from a volunteering subject.

S3.3 LEARNING OBJECTIVES

The objectives of this studio are to:

- Understand small signal rectification using analog circuits.
- Implement an active filter.
- Understand the advantages of pre-processing the physiological signals before sampling into a computer.

S3.4 SAFETY NOTES

It is extremely important that the subject is not connected directly to the ground of any of the instruments used during this studio – e.g., oscilloscope, voltage supply, computer. In hospitals the clinical equipment has isolated grounds that prevent the passage of alternating (AC) or or direct (DC) current from such equipment to the patients. Isolated grounds are a safety measure to protect patients in case of faulty equipment. In the worst situation, a direct connection between the subject and the ground of a faulty apparatus may result in a fatality. All electric equipment in the laboratory should be tested for safety prior to usage. This is normally the responsibility of the host institution.

S3.5 EQUIPMENT, TOOLS, ELECTRONIC COMPONENTS AND SOFTWARE

Additional information about the use of the items required in this studio can be found in Studio 1 and in the Appendices.

Equipment

a. Breadboard.

b. Power supply (+15V, –15V).

c. Multimeter.

d. Oscilloscope.

e. Grass IP511 AC Amplifier.

f. 3 Lead cable for the Grass Amplifier.

g. EMG or ECG electrodes.

h. Conductive ECG electrode gel (optional).

Tools

a. BNC-to-microclip cables.

b. Jumper wires for the breadboard.

Electronic Components (remember to download the available datasheets)

a. 2 1N4148 or similar small signal diodes.

b. 1 Quad-Operational Amplifier, TL084CP.

c. Various 1/4W resistors and capacitors (low leakage preferred).

1- 1K.

2 - 91K.

1–0.15 μF.

1–0.33 μF.

2– 100 nF K (bypass capacitors).

d. 22 pin header (optional).

1–3 pin header (optional).

Software

- MATLAB®.

S3.6 CIRCUIT OPERATION

The circuit consists of two parts (Figure 3.1). U1A and U1B form the absolute value or full-wave rectifier circuit and the other Op-Amp section U1D forms a two-pole Sallen-Key Butterworth low-pass filter with a corner frequency of 8 Hz. The absolute value amplifier always provides a positive going output for either a negative or positive input. For a positive input, diode D1 conducts and D2 is open. The output of U1B is fed back to both

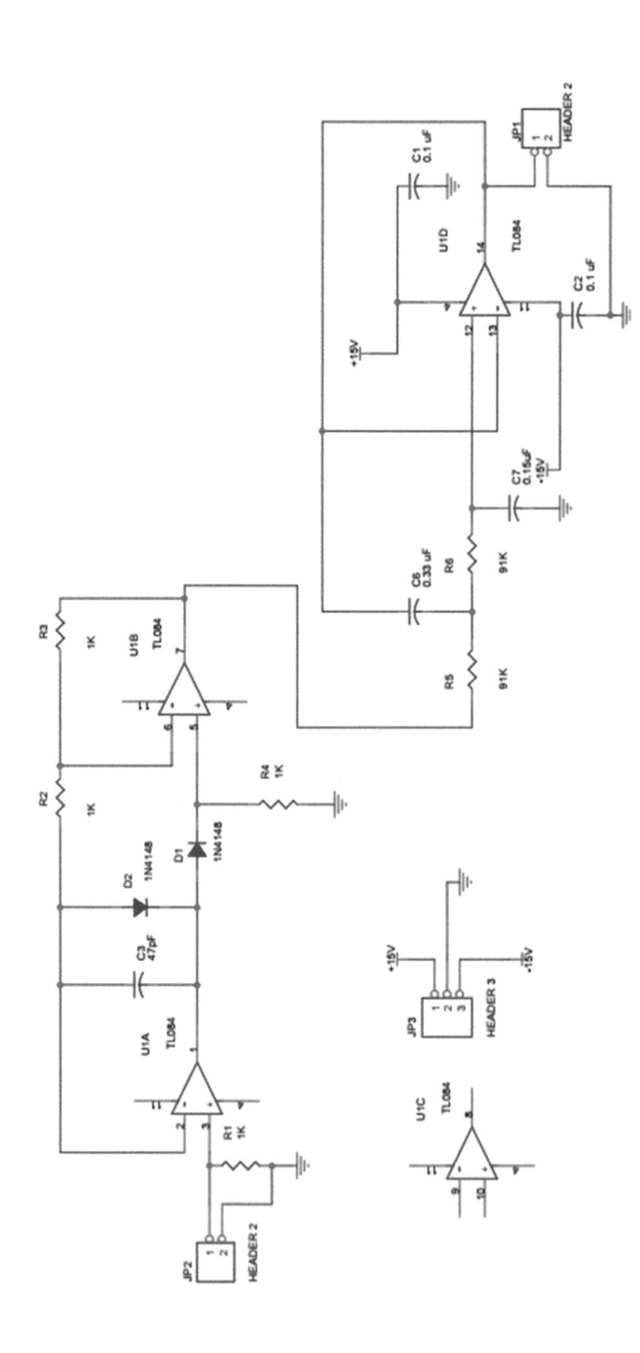

FIGURE 3.1 Schematic diagram of small signal rectifier and low-pass filter.

Op-Amp sections negative inputs through R3 and R2. Since the loop is closed with a direct connection to the inverting inputs with no path for current to flow, the gain is unity. U1A's DC output will be positive by one diode drop (~0.65 volts) from the output signal's amplitude. Diode D2 is not conducting since the voltage is essentially the same on both sides of it. For negative going input signals, D1 is reversed biased so U1B's positive input is at ground, and current flows through R3, R2 and D2 so that the input voltage is applied at unity gain to the left side of R2. In this case, U1A's output will be one diode drop below the signal at the input. U1B then becomes an inverting amplifier with a gain of –1, making the output again positive, hence providing the absolute value. You can calculate the exact value of f_c with the following formula:

$$f_c = 1/2\pi\sqrt{R5 \times R6 \times C6 \times C7}$$

The Q of the circuit can be calculated using the formula:

$$Q = \sqrt{mn}\,/\,(m + 1 + mn(1 - K)$$

Where: m = R5/R6 = 1

N = C6/C7

K = 1 (gain)

The equation then simplifies to $Q = \sqrt{n}\,/\,2$.

S3.7 DETAILED EXPERIMENTAL PROCEDURE

1. Build the circuit given in the schematic on a breadboard. Temporarily leave out R2 and D1, but leave room for them. Also, temporarily jumper U1.3 to U1.5.

2. Compare the circuit with the schematic and verify all the connections before applying the power supply.

3. Apply DC power to the breadboard at JP3. Apply a 2 Hz sinusoidal signal with ±1 V (peak-to-peak) amplitude from the signal generator to the input at JP2 of the circuit (with respect to ground).

4. Now observe the output of the last op-amp at JP1 on the oscilloscope (Channel 1) as well as the original sinusoidal signal from the signal generator (Channel 2) and compare if the waveforms are identical, if

there are any distortions in the output waveform, and if there is any phase shift.

5. Test your circuit at frequencies up to 100 Hz in 2 Hz steps to 20 Hz, and then in 10 Hz steps while making amplitude measurements on the oscilloscope.
6. Make a Bode plot of the output amplitude and phase as a function of frequency in Excel or MATLAB. Use log-log scale for the axis.
7. After confirming the low-pass filter is operating properly, turn off the power remove the temporary jumper and install R2 and D1. Turn the power back on.
8. Set your signal generator back to 2 Hz. The output on channel 1 of the scope should be the absolute value of the input sine wave, as in Figure 3.2.

   NOTE: If the signal generator output has a DC offset, or the gains for positive and negative cycles of the input signal are different in your circuit, the output will have unequal amplitudes for each half cycle of the sinusoidal signal.

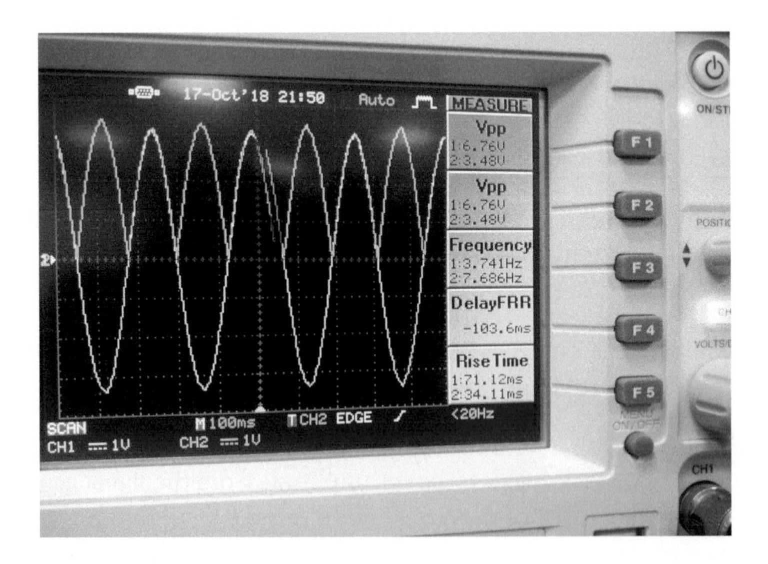

FIGURE 3.2 Output of the full-wave rectifier with equal amplification for both positive and negative inputs. Blue trace is the output; yellow trace is the sinusoidal input.

9. At input frequencies around the corner frequency of the low-pass filter (8 Hz), the output will be a sinusoidal running on top of a DC voltage, that height of which is proportional with the amplitude the input signal (see Figure 3.3).

   If you increase the generator frequency to 100 Hz, you should have a DC output that is changing only by the amplitude of the input signal (see Figure 3.4). Why does the output become a DC level that is independent of the input frequency at frequencies much higher than the filter corner frequency?

10. Now, disconnect the signal generator and hook the cable to the BNC output connector on the Grass amplifier. Get an input cable for the Grass amplifier and connect it to one of the subjects in the group placing three electrodes on the subject's arm with the ground lead below the wrist opposite the index finger. Place another electrode V+ just behind the first one but rotated on the arm to be opposite the pointer finger. Place the V− electrode just below the elbow.

11. Use sufficient amplification and appropriate filtering for EMG (10–500 Hz) to see a clear EMG waveform on the oscilloscope. Generate the EMG signal by making a fist and squeezing it.

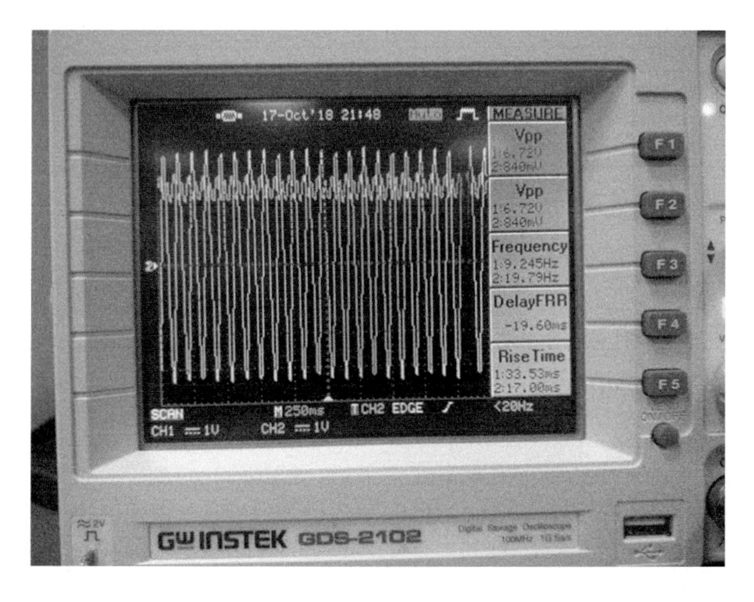

FIGURE 3.3 The output of the low-pass filter for a sinusoidal input at a frequency slightly higher than the corner frequency of the filter (8 Hz).

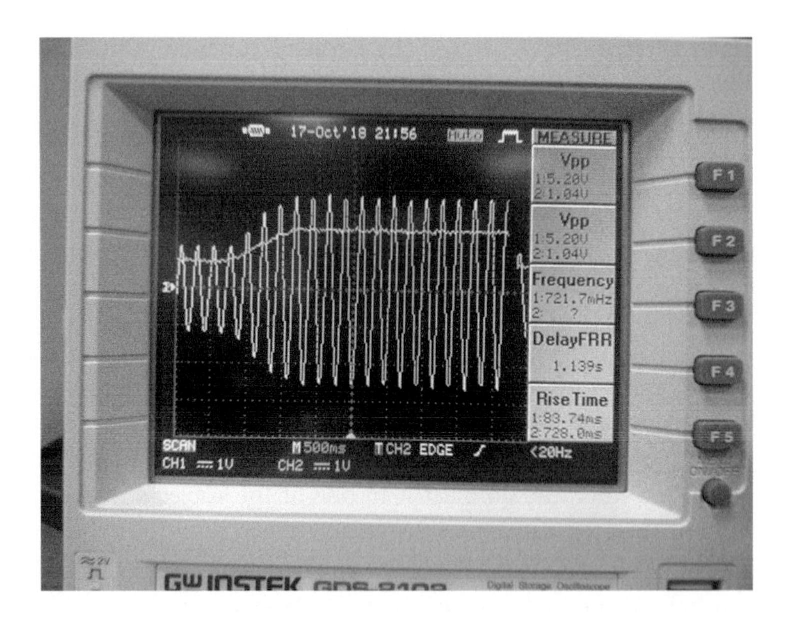

FIGURE 3.4 The output of the low pass filter for a sinusoidal input at a frequency much higher than the corner frequency of the filter. The sinusoidal input amplitude is increased manually during the first 1.5 s.

12. Do you see the raw EMG signal and the rectified one on the scope? Does the rectified signal follow the outline of the EMG envelope? Is there a delay? (see Figure 3.5)

In your report, remember to include:

1. Introduction: theory on Sallen-Key low pass filters and absolute value circuits.
2. Results: plot the amplitude and phase of the transfer function using a log scale for horizontal axis (frequency) and units of dB for the amplitude.
3. Discussion: discuss the problems encountered and how they were solved in the building and testing of the circuit. Discuss potential usages of the circuit for electrophysiological signals. Discuss limitations of the circuit in terms of frequency, noise and signal amplitude.
4. Discuss the usage of this circuit in EMG analysis.

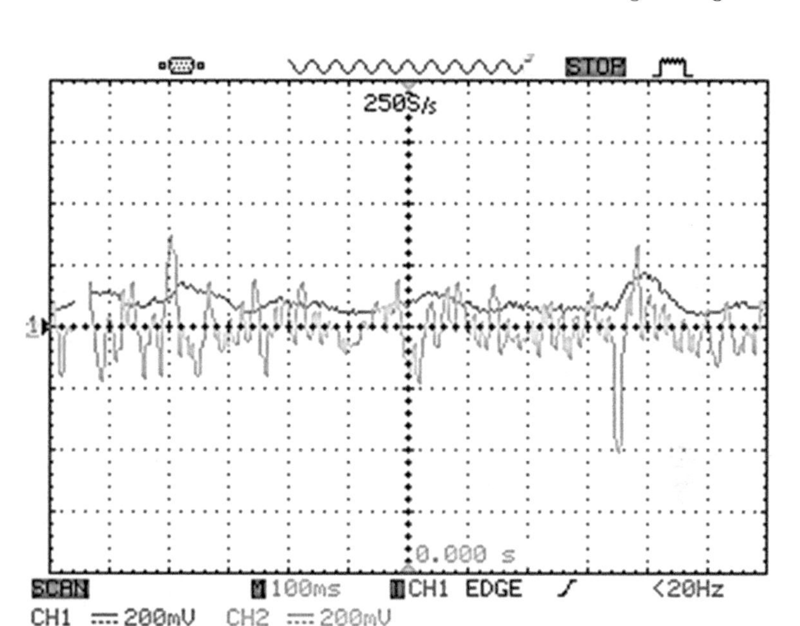

FIGURE 3.5 Snapshot of the rectified-filtered output signal (smoother CH1) during the acquisition of small-amplitude EMG signals from a subject's arm muscle. The input to the circuit coming from the EMG amplifier is CH2.

S3.8 CIRCUIT TESTING AND TROUBLESHOOTING

1. Before turning the power on, carefully check if the voltage supply is applied to the circuit with the correct polarity. Using a voltmeter with one lead connected to the ground, check the +/– 15 V and ground voltages at the power terminals of the Op-Amp. It is a time saving practice to do this as the first troubleshooting step in any circuit, analog or digital.
2. Another good practice is to measure the total current that the circuit is drawing from the power supply (most modern power supplies have a voltage and a current display). If the current is higher than a few tens of a mA, or one of the chips is becoming too warm (be careful not to burn your fingers) you may conclude that there is a broken chip or a short circuit due to an incorrect connection.

S3.9 QUESTIONS FOR BRAINSTORMING

a. Why was a Butterworth filter selected?

b. What happens if you increase the filter cutoff frequency?

c. What happens if you make the filter cutoff frequency too low?

S3.10 IMPORTANT TOPICS TO INCLUDE IN THE LAB REPORT

a. The applications for rectification and filtering with EMG signals.

REFERENCES AND MATERIALS FOR FURTHER READING

References marked with an asterisk (*) are recommended to those interested in expanding on the content of this chapter.

1. *TL08xx Datasheet.* Texas Instruments, SLOS081I, May 2015.
2. *Gans, Carl; Loeb, Gerald E., *Electromyography for Experimentalists.* University of Chicago Press, Chicago, 1986.
3. Kamen, Gary; Gabriel, David, *Essentials of Electromyography.* Human Kinetics Publishers, 2009.
4. **Analysis of the Sallen-Key Architecture.* Texas Instruments, SLOA024B, September 2002.

Studio 4

Digital Voltmeter

Usage of Analog-to-Digital Converters

S4.1 BACKGROUND

An analog-to-digital converter (A/D or ADC) converts an analog voltage to a digital number.

The AD7575 A/D used in this studio is a successive approximation A/D. The first stage at the input consists of a sample and hold circuit that samples the incoming analog signal and holds its value constant for the entire A/D conversion cycle. The output of the sample and hold connects to a comparator. The second input to the comparator is from a digital to analog converter (D/A or DAC) that has the same number of bits as the resolution of the A/D. The D/A is driven by a successive approximation register (SAR). On the first comparison, this register is set with the most significant bit (MSB) at 1 and the other bits at 0. If the comparator determines that the input signal is higher than the voltage corresponding to the digital number when the MSB is set high, the SAR leaves the MSB high and moves to the next bit. Contrarily, if the input is less than DAC output, the SAR sets the MSB to 0, and then moves to the MSB-1 bit. This process is repeated for all the bits in the SAR. When the conversion is completed, the SAR output gets reported as the output of the A/D.

The Schmitt trigger, named after the American inventor Otto Schmitt, is a comparator that has hysteresis at its input. A comparator is a device that compares two inputs, and reports which input is larger at the output.

Hysteresis is created by adding feedback from the output to the positive input so that the switching threshold at the input changes when the output state changes in a manner that creates a positive shift in the switching threshold when the output goes positive, and a negative shift when the output goes low. This prevents instabilities from occurring as the signals move through the threshold. The 74HC14 integrated circuit (IC) used in this experiment is an inverting buffer with hysteresis at its input, i.e., inverting Schmitt trigger.

This studio uses a seven-segment LED display as its output. This display can show the digits 0– 9. To drive this display, we will use a decoder (CD4511) that converts the binary coded decimal (BCD) data to turn on the required segments of the display such that the number displayed corresponds to the digital value of the A/C output. The CD4511 BCD to seven-segment driver uses combinatorial logic to decode the information to drive the display. The same logic could be implemented in a look-up table. The logic used is defined in a "Truth Table."

S4.2 OVERVIEW OF THE EXPERIMENT

This laboratory exercise will introduce the concept of analog-to-digital (A/D or ADC) converters and give a practical example of a specific A/D converter circuit built around AD7575. This particular A/D converter is primarily manufactured to be used as a peripheral to a microprocessor or a microcontroller. The control inputs, however, are simple enough that they can be emulated with a simple digital logic circuit. The output will be displayed on an LED display as opposed to be placed on the data bus to be read by the microprocessor. The third block in the circuit is a square wave generator that is built around an inverting Schmitt trigger gate (74HC14). This simple circuit will introduce the basic concept of a resistor-capacitor (RC) oscillator that is based on charging and discharging of a capacitor through a resistor. Overall, the circuit has three blocks; the A/D converter, the display decoder (CD4511) and the LED display, and the RC square wave generator.

S4.3 LEARNING OBJECTIVES

The students will:

1. Learn the usage of analog-to-digital converters.

2. Become familiar with seven-segment LED display decoders and displays.

3. Learn to build a simple square-wave oscillator using a Schmitt trigger.

S4.4 NOTES ON SAFETY

There is no health risk in this studio other than standard precautions that need to be observed in dealing with electronics. No human subject is required.

S4.5 LIST OF MATERIALS

Electronic Components:

a. Breadboard (protoboard).

b. A/D converter AD7575JNZ.

c. Seven-segment common-cathode LED display decoder (CD4511BE).

d. Seven-segment common-cathode LED display (HDSP-513A).

e. Schmitt trigger (SN74HC14N).

f. 10 kΩ trimpot (10-turn is preferred) (Bourns PV36W103C01B00).

g. A 1.2 V voltage reference (e.g. LM385BLP-1-2) or equivalent.

h. Various ¼ W resistors and capacitors (low leakage preferred):

   2.2 K.

   10 K.

   68 K.

   100 K.

   470.

   100 pF.

   10 nF.

   100 nF.

i. BNC-to-micro clips cables.

Equipment:

a. Single voltage supply (5 V).

b. Multimeter (with capacitor meter option if available).

c. Oscilloscope.

S4.6 CIRCUIT OPERATION

a. Analog-to-Digital Converter:

The integrated circuit (IC) chosen here, AD7575 has three main input control terminals, a voltage reference input, and the voltage to be measured (Ain). The conversion cycle begins when both the chip select (CS) and the Read (RD) control inputs are lowered to *Logic 0.* While the instantaneous value of Ain is being converted to an eight-bit digital number, the chip puts out the digital value calculated in the previous cycle on the eight-bit data bus for the microprocessor (or it is CD4511 in our case) to read the value. The busy output goes low indicating that the number on the data bus is a valid number that can be accessed. One can read the previous cycle results (old number) by detecting the falling edge of the "busy" signal. Alternatively, the result of the current cycle is available at the rising edge of the "busy" output which then can be used to latch this number into the display decoder, as it is done by connecting the "busy" output to the latch enable (LE) input of the decoder in this circuit. Reading the old or the new number virtually does not make any difference since the conversion cycle is very short (i.e., one cycle of the Schmitt trigger oscillator) and the results would be shifted only by one cycle at the display output. Note that only the most significant four bits of the A/D output are latched to the decoder and the lower four bits are ignored due to the fact that we have only one display decoder and one digit display in this circuit. Simply not reading the lower half of the eight-bit output lowers the resolution of the converter from eight to four bits.

b. Display Decoder and Display:

The four-bit number on the DB4 through DB7 pins of the data bus is latched into the decoder on the rising edge of "busy" signal. The CD4511 converts the four-bit input sequence into a seven-bit sequence to turn *on* or *off* individual segments of the display. All seven LEDs of the segments and the decimal points are connected together on their cathodic terminals, which forms the common ground terminal of the display, hence the name common-cathode display. If a particular CD4511 output is *Logic 1*, a current determined by the value of the series resistor flows through the LED of the segment into the ground. A *logic 0* at the output turns the segment off. For instance, if all four-bit inputs to CD4511 are zero, all outputs a through f would be *Logic 1* and the output g (middle segment) would be *Logic 0*, thus displaying a zero character. Figure 4.1 is the truth table for the CD4511. Note that the display does not show any of the hexadecimal characters A through F (1010 through 1111) but becomes blank. The CD4511 could be replaced by a look up table that has the input to the memory used

TRUTH TABLE

LE	$\overline{BI}$	$\overline{LT}$	D	C	B	A	a	b	c	d	e	f	g	Display
X	X	0	X	X	X	X	1	1	1	1	1	1	1	8
X	0	1	X	X	X	X	0	0	0	0	0	0	0	Blank
0	1	1	0	0	0	0	1	1	1	1	1	1	0	0
0	1	1	0	0	0	1	0	1	1	0	0	0	0	1
0	1	1	0	0	1	0	1	1	0	1	1	0	1	2
0	1	1	0	0	1	1	1	1	1	1	0	0	1	3
0	1	1	0	1	0	0	0	1	1	0	0	1	1	4
0	1	1	0	1	0	1	1	0	1	1	0	1	1	5
0	1	1	0	1	1	0	0	0	1	1	1	1	1	6
0	1	1	0	1	1	1	1	1	1	0	0	0	0	7
0	1	1	1	0	0	0	1	1	1	1	1	1	1	8
0	1	1	1	0	0	1	1	1	1	0	0	1	1	9
0	1	1	1	0	1	0	0	0	0	0	0	0	0	Blank
0	1	1	1	0	1	1	0	0	0	0	0	0	0	Blank
0	1	1	1	1	0	0	0	0	0	0	0	0	0	Blank
0	1	1	1	1	0	1	0	0	0	0	0	0	0	Blank
0	1	1	1	1	1	0	0	0	0	0	0	0	0	Blank
0	1	1	1	1	1	1	0	0	0	0	0	0	0	Blank
1	1	1	X	X	X	X				*				*

X ≡ Don't Care * Depends on BCD code previously applied when LE = 0

Note: Display is blank for all illegal input codes (BCD > 1001).

FIGURE 4.1 CD4511 Truth Table (4.11 C).

as its address and the output would be the data from the memory to drive the display, although a buffer/driver would probably be needed between the memory and the display which is integrated into the CD4511.

The Avago HDSP513A display used is a red LED display that has seven segments and a decimal point with a common cathode configuration. Each segment can be driven with up to 15 mA of current if continuously enabled as in our application. When driven at 10 mA, the LED forward voltage is typically 2.06 volts. The CD4511 drive voltage will be about 3.9 volts with a 10 mA load current. The drive current in our application is about 4 mA (that is (3.9–2.06 V)/470 Ω) using a 470 Ω current limiting resistor in series with each segment.

c. Square Wave Oscillator:

This circuit makes use of the hysteresis of Schmitt triggers as shown in the input-output plot of Figure 4.2. Hysteresis means the output transitions from low-to-high and high-to-low at different values of the input voltage (V_T- and V_T+), typically around one third and two thirds of the power supply voltage. Initially, when the power supply is turned on in Figure 4.3, the input capacitor is discharged (zero volts). The output of the gate is at *Logic 1* or power supply voltage because of the inverting function. Thus, the capacitor begins to charge through the resistor since the right side of the resistor is at high and the left terminal at low voltage. As the capacitor

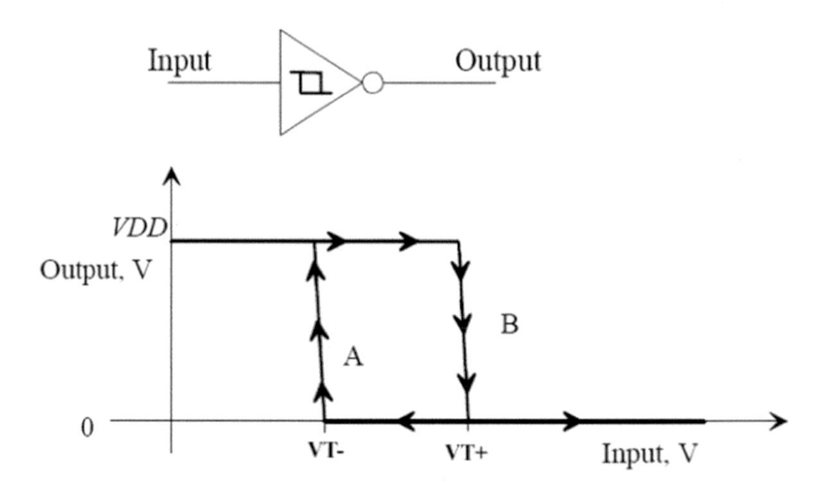

FIGURE 4.2 Hysteresis property of an inverting Schmitt trigger.

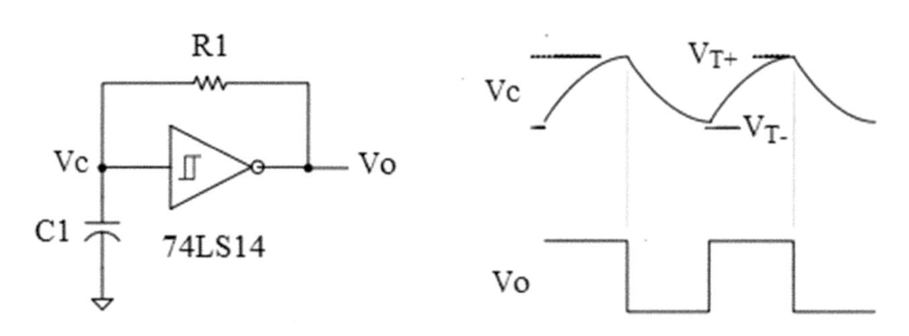

FIGURE 4.3 Input and output waveforms of the Schmitt trigger oscillator.

charging and the input voltage is increasing from zero, the output remains at positive power supply voltage. At the time instant the capacitor (i.e., input) voltage crosses two thirds of the power supply voltage, the output switches to ground potential since the input now is considered at *Logic 1*. Now the right side of the resistor is at ground and the left side is at two thirds of the power supply. Thus, the capacitor begins to discharge through the resistor with a current flowing in the reverse direction. The capacitor needs to reach all the way down to one third of the power supply before the output can switch back to high voltage, due to hysteresis. At that point, the resistor current switches again since the output goes high and the input is at one third of the power supply. The capacitor starts charging again through the resistor. This charging/discharging cycle of the capacitor between one third and two thirds of the power supply voltage continues indefinitely, producing a 50% duty cycle square wave at the output. The frequency of this oscillator is defined by the RC values according to the following equation:

$$f = 1/(0.69 * R * C)$$

S4.7 DETAILED EXPERIMENTAL PROCEDURE

1. Build the circuit shown in Figure 4.4 on a breadboard. A good practice while building a circuit of this size is to highlight each connection on a printout of the schematic as they are transferred to the breadboard. One can easily detect if any of the connections are missing at the end if the highlighting is done carefully.

2. Notice that the circuit needs only one 5 V power supply. The A/D converter and the display decoder can handle higher voltages, but this specific Schmitt trigger (HC-high speed CMOS) cannot exceed 6 V

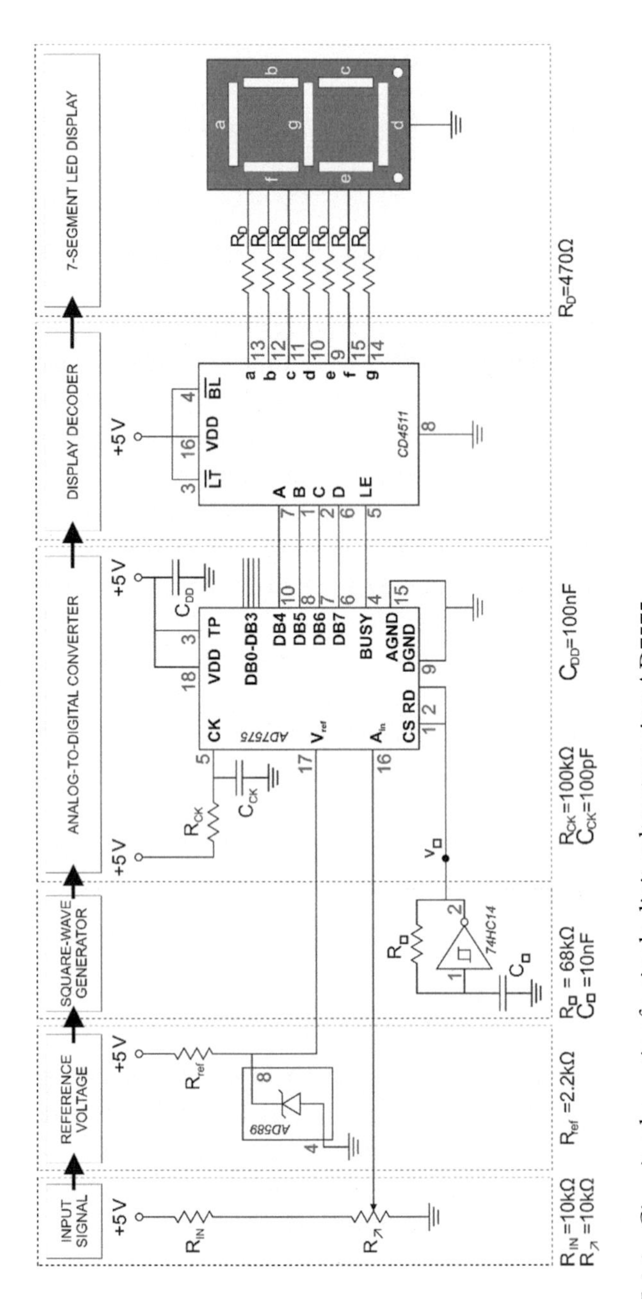

FIGURE 4.4 Circuit schematic of a single digit voltmeter using AD7575.

maximum, as recommended by the manufacturers. Thus, the voltage supply should be adjusted carefully before connecting to the circuit. A good practice is to turn the power supply off after adjusting the voltage output, make the connections to the circuit (for instance using banana-to-micro clip cables), and then turn the power supply on.

3. The display should change as the input voltage (which is the voltage being measured by this voltmeter) is varied by turning the potentiometer at pin 15 of the A/D converter. If not, follow the troubleshooting steps below before continuing with the experimental procedure.

4. After troubleshooting the circuit, turn the input voltage all the way down to zero and begin increasing slowly. Make a table of voltage values at the transition points of the display from one number to the next. If the reference voltage is 1.23 V the display should show 1 for 1 V, 2 for 2 V …, etc. up to 9 V. At 10 V it should go blank since CD4511 cannot convert this value to a meaningful number. (Note: hexadecimal displays show A, B, C, D, E and F for numbers between ten and 15.)

5. Now change the reference voltage to see its effect on the transition points of the display. You may add another LM385 in series to the first one to double the reference voltage. Because the input voltage is converted using this reference voltage and assumes that it is 1.23 V, for a V_{ref} equals 2.46 V, the output transitions will take place at double the values. So, 2 V should now become 1 on the display, and 18 V should become 9 V, etc.

6. Do all the possible mistakes that are discussed below in the last item of troubleshooting intentionally and observe the results. Do the following *one at a time*. Discuss the source of the problem with your classmates:

   a. Disconnect the ground terminal of seven-segment display (note it suffices to connect one of the ground pins of the display to the ground for normal operation).

   b. Switch or disconnect some of the seven connections from the display decoder to the display.

   c. Switch some of the data bus pins (DB4 through DB7) between the A/D converter and the display decoder, observe the display while varying the input voltage, Ain.

7. In order to better understand the function of the decoder control inputs do the following *one at a time* and observe what happens to the display. Look up CD4511's datasheet to confirm your findings:
   a. Connect the light (LT) input to ground and observe the display.
   b. Connect the blank (BL) input to ground and observe the display.
   c. Connect the latch (LE) input to 5 V power and change Ain. What happens to the display?

S4.8 CIRCUIT TESTING AND TROUBLESHOOTING

- Before turning the power on, carefully check if the voltage supply is applied to the circuit with the correct polarity. Using a voltmeter, one lead connected to the ground, check the 5 V and ground voltages at the power terminals of all chips. It is time saving practice to do this as the first step of troubleshooting in any circuit, analog or digital.
- Another good practice is to measure the total current that the circuit is withdrawing from the power supply (most modern power supplies have a voltage and a current display). If the current is higher than a few tens of a mA, or one of the chips is becoming too warm (be careful not to burn your fingers) you may conclude a broken chip or a short circuit due to a false connection.
- Check if the square wave generator is working properly by observing its output (pin 2 of 74HC14) on the oscilloscope. The frequency should be around 1 kHz as determined by the RC components, and the voltage should be switching between zero and 5 V. Some jittering of the waveforms is expected because of the dependency of the RC values on temperature and threshold voltages on the power voltage. Another 100 nF capacitance can be placed near this chip between 5 V and ground to suppress any glitches and noise in the power supply. The function of this circuit block is to control the conversion cycle of the A/D converter and thus without a proper square wave output the converter will not produce digital outputs.
- Next, check if there is a proper high frequency (~1 MHz) characteristic waveform at pin 5 of AD7575 using the oscilloscope. This shows that the internal clock that controls individual events within a single conversion cycle is functioning properly. A sample waveform is shown in Figure 4.5 bottom plot.

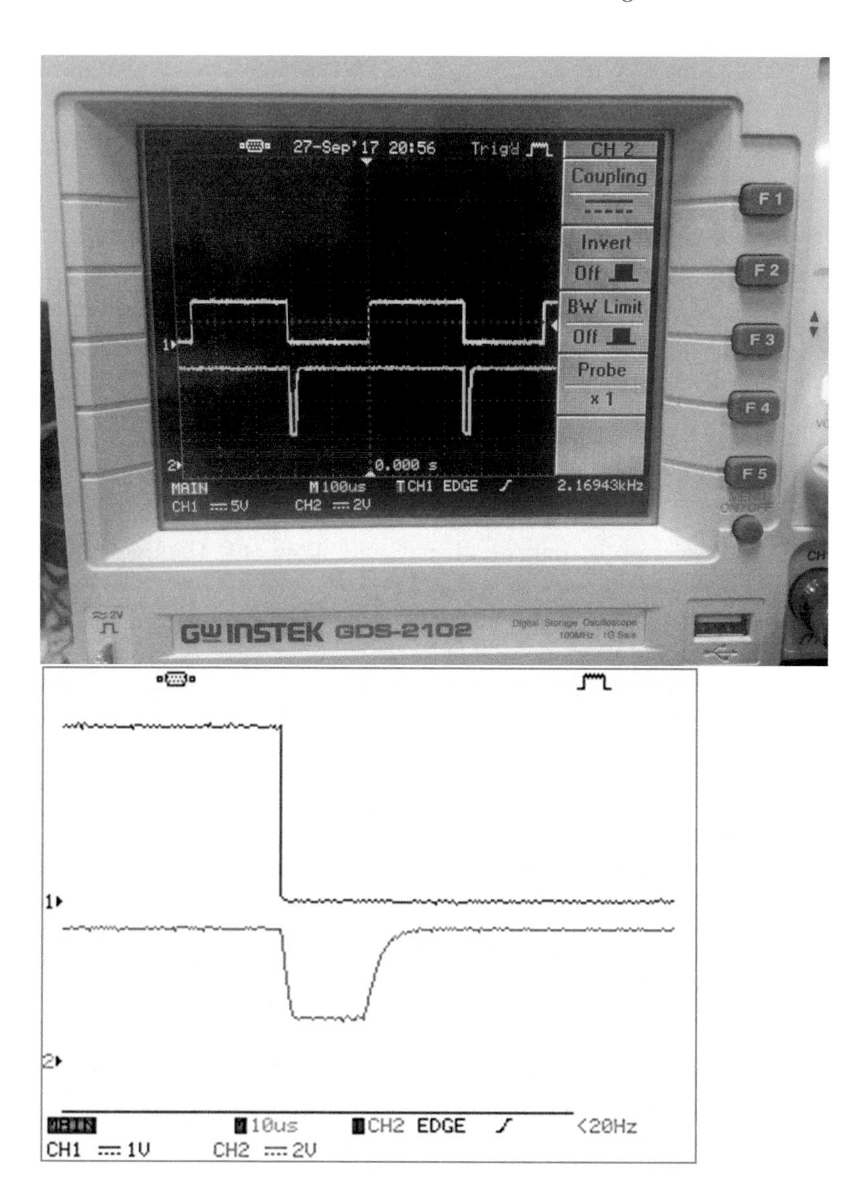

FIGURE 4.5 Top: A/D busy signal at pin 4 (blue) and chip select (CS) at pin 1 (yellow). Bottom: The CS (yellow) and the clock (CLK) signal at pin 5 of AD7575 in an expanded time scale. Note that the CLK signal is only a few microseconds long and does not reach zero volts at the low level.

- Next, using a voltmeter check if the reference voltage (V_{ref}) at pin 17 is correct (1.24 V if the LM385 is used, AD589 is an alternative that has a typical reference voltage of 1.235 V). A common mistake is to connect the reference component in the reverse direction.
- Finally, connect a voltmeter to the Ain input (pin 16) using alligator clips or wires and watch the voltage change as the 10 kΩ potentiometer is adjusted. It is best to use a 10-turn trimpot or potentiometer here for fine adjustment of the voltage from zero up to 2.5 V.
    - If the numbers are increasing and decreasing in a non-sequential manner as the input voltage (Ain) is being varied, this indicates a connection problem from A/D converter data bus to the display decoder. They may be switched or left open. A common observation is that the display will show only the odd or even number (DB4 is connected to the ground or power) or it will flicker continuously if one of the data inputs to CD4511 is left open.

S4.9 QUESTIONS FOR BRAINSTORMING

a. What is the resolution of your voltmeter?

b. How would you change the input range (and hence the resolution) of this voltmeter so that 0.1 V at the input reads as 1, 0.2 V reads as 2, etc.? Note that you can turn on the decimal point LED on the display by connecting the corresponding terminal to positive power through a resistor.

c. How would you increase the number of digits so that for instance 1.1 V reads as 1.1? How many bits are needed for a two digit display?

d. What happens if you lower the square wave frequency down to 10 Hz generated by 74HC14?

e. What happens if the R or C value of the square wave generator is increased?

f. What happens if the resistor values from CD4511 to the display are halved in value?

S4.10 IMPORTANT TOPICS TO INCLUDE IN THE LAB REPORT

a. Explain the timing of the control inputs and the outputs of AD7575 using the "timing diagram" from its data sheet.

b. Study the input-output conversion formula of AD7575 given in its data sheet, which includes Vref.

c. Define the difference between common-anode and common-cathode LED displays.

d. Explain the "Truth Table" given in CD4511 data sheet.

e. Explain the input-output relation of a Schmitt trigger in comparison to a standard buffer or inverting gate.

f. Describe any troubleshooting you had to do during this laboratory and how you fixed the problem.

REFERENCES AND MATERIAL FOR FURTHER READING

1. Kosonocky, S.; Xiao, P. Analog-to-digital conversion architectures, Chapter 5 in *Digital Signal Processing Handbook*, eds. Vijay K. Madisetti and Douglas B. Williams. CRC Press LLC, Boca Raton, 1999. http://dsp-book.narod.ru/DSPMW/05.PDF.
2. *AD7575 Datasheet*, Analog Devices Revision B. July 1998, Norwood, MA.
3. *CD4511B Datasheet*, Texas Instruments, SCHS072B, July 2003, Dallas, TX.
4. *HDSP-513 Datasheet*, Avago, AVO2-1363EN, June 21, 2008.
5. *SN74HC14N Datasheet*, Texas Instruments, SCLS085J, October 2016.
6. *LM385-1.2 Datasheet*, Texas Instruments, SLVS075J, January 2015.

STUDIO 5

Force Measurements with PZT Transducers

S5.1 BACKGROUND

Piezoelectric devices produce an electric charge when under mechanical stress. Lead zirconate titanate (PZT) is a widely used piezoelectric ceramic with a perovskite crystalline structure. When the PZT is baked at a high temperature, a crystalline structure is formed. The piezoelectric effect generates an electric charge when the material is deformed. Conversely, when an electric potential is applied between the facets of the material, it deforms.

S5.2 OVERVIEW OF THE EXPERIMENT

The objective of this experiment is to observe the piezoelectric effect. A charge amplifier is assembled to produce an output voltage that is proportional with the total charge generated by a PZT transducer. The output of the amplifier will be analyzed in MATLAB®. The mechanical resonance will be demonstrated using the frequency characteristics of the transducer (Figure 5.1).

S5.3 LEARNING OBJECTIVES

The objectives of this studio are to:

- Understand the operation of the piezoelectric materials as a force sensor.

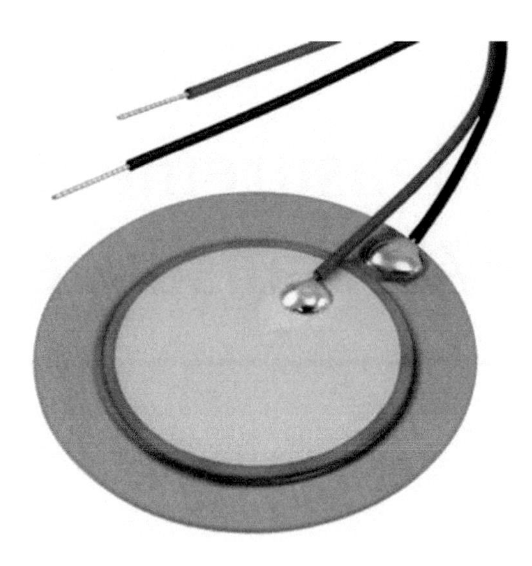

FIGURE 5.1 PZT transducer (Murata 7BB-20-6L0).

- Understand the mechanical resonance of the piezoelectric sensors in frequency domain.
- Understand charge amplifiers.

S5.4 EQUIPMENT, TOOLS, ELECTRONIC COMPONENTS AND SOFTWARE

Additional information about the use of the items required in this studio can be found in Studio 1 and in the Appendices.

Equipment:

- Power supply (±9 V).
- Multimeter.
- Oscilloscope.

Tools:

- BNC-to-microclip cables.
- Banana plug to microclip cables.

Components:

a. Murata 7BB-20-6L0 PZT transducer.
b. Breadboard (protoboard).

b. Operational Amplifier TL082ACP or similar.

c. Various ¼ W 5% resistors and capacitors (C1 should be low leakage).

   1–6.8 MΩ.

   1–0.22 μF.

   1–0.047 μF.

   2–0.1 μF.

e. 1-2 pin header (optional).

f. 1-3 pin header (optional).

Software:

- MATLAB®.

S5.5 DETAILED EXPERIMENTAL PROCEDURE

1. Take a piece of electric tape and tape the PZT transducer (Murata, part no. 7BB-20-6L0) on a clean, flat, hard surface (like the lab bench) with the brass side facing the bench. Apply force on the PZT with a finger. The surface should be perfectly flat and there should not be any particles under the PZT which may cause the transducer to break. Tape the PZT with a piece of electric tape to the bench to make sure that there is no electrical contact with the finger while keeping it mechanically stable.

2. Build a charge amplifier in Figure 5.2 using an Op-Amp (TL082), a 220 nF capacitor in the feedback loop. Attach the red PZT lead to the inverting input of the amplifier and the black one to the ground. Push a separate piece of solid wire into the same hole with the PZT wire to make sure that they are tightly inserted into the holes of the breadboard (PZT leads are too thin for breadboard holes) (Figure 5.2).

3. Connect the output of the circuit to the oscilloscope using a micro-clip to BNC cable.

4. Power the Op-Amp with a +/– 9 V supply.

5. Apply a medium force with a finger on the PZT over the part covered with electric tape. Be careful not to touch the solder sites. Try to maintain the force at the same level with your finger after the initial

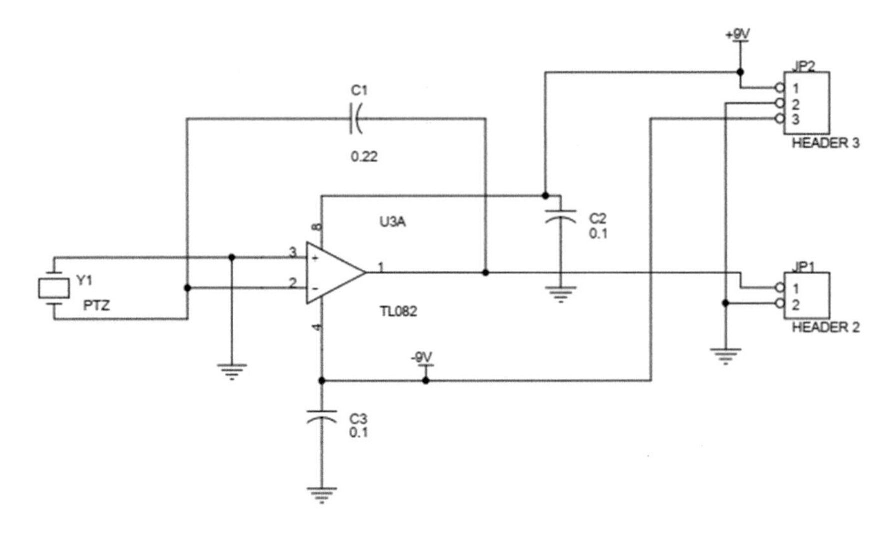

FIGURE 5.2 Charge amplifier.

application. A better technique is to use a metal object that weighs about 100–200 g to obtain reproducible results.

6. Observe the waveform on the oscilloscope. The voltage should stay flat for a constant force, after the initial oscillation in the signal due to placement of the weight. With the force being constant no additional charges are produced by the PZT, and the circuit output is proportional with the total charge that has been collected from the PZT up to that point in time (Figure 5.3).

7. What happens if you add a 6.8 MΩ parallel resistor to the feedback capacitor? The fact that the voltage returns to the baseline indicates that the capacitor is discharged through the resistor if no additional charges are coming from the PZT (Figure 5.4).

8. What happens if you take the capacitor out of the circuit and leave the resistor in? In this case the circuit behaves like a differentiator of the force input. The output changes only when there is a change in the force. At the onset of the force, the output produces a positive voltage spike, and at the offset of the force a negative voltage spike.

9. Place the capacitor back in the circuit and remove the resistor. What would happen if you switch the black and red wires from the PZT to the amplifier? The output polarity should change if you switch the PZT leads.

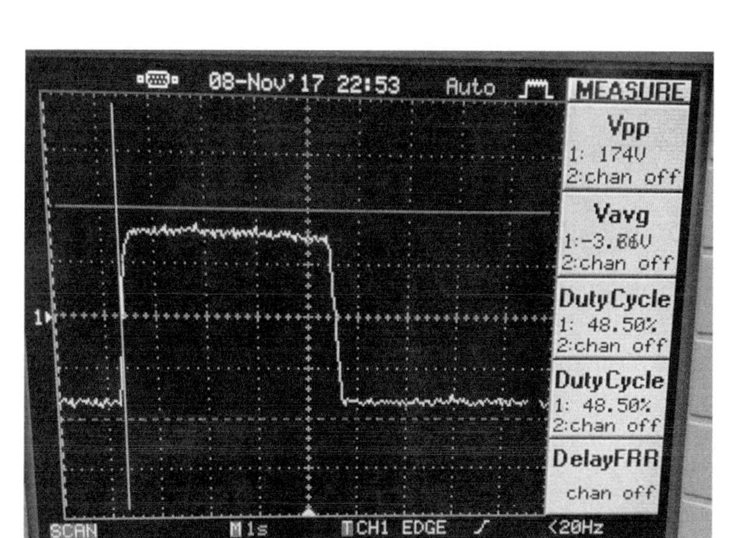

FIGURE 5.3 The output voltage during loading and unloading of a 200 g weight on the sensor. The output step change is about 1.74 V. (The oscilloscope assumes a probe gain of ×100 which is actually 1. This is a common mistake. Check the assumed probe gain from the oscilloscope screen menu). Capacitor is 22 nF. This corresponds to a k value of 19.5 nC/N according to Equation 5.1.

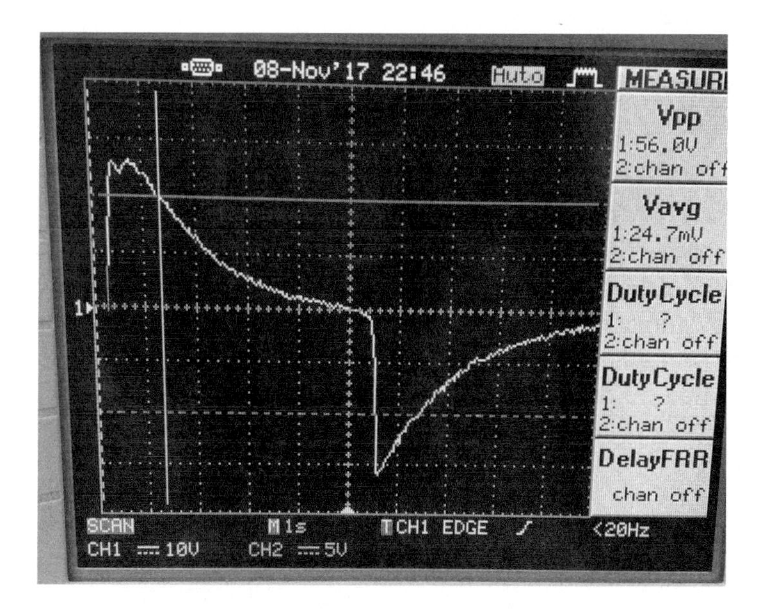

FIGURE 5.4 The output waveform when a 6.8 MΩ resistor is placed in parallel to the capacitor, causing a slow discharge of the capacitor, both during loading and unloading.

10. Now, change the capacitor in the charge amplifier to 47 nF. Leave the parallel resistor out of the circuit. Apply a similar force to the PZT and observe it on the oscilloscope. Is the signal larger this time?

11. Acquire all signals into MATLAB for your report and further analysis. Use "*softscope*" (or "*analogInputRecorder*" for newer version of MATLAB) tool and follow these instructions in softscope:

    a. In the hardware setting box, chose "Dev 1" and channel 0 as input, 100 samples/sec for the sampling rate, and "Single Ended" for the input configuration. Hit OK.

    b. Hit *Trigger* button to start viewing the signal and hit *Stop* to stop it.

    c. To save the data you see on the display go to *File* menu and choose *Export > Channels*. Select *Workspace (array)* option and hit *Export* to export the data to the workspace of MATLAB. You can define the number of samples to be saved as well.

    d. Go to the Command window and type *save filename c0/ascii* to save the first channel data to the hard disc.

12. With the 220 nF capacitor (no parallel resistor), apply a known weight (e.g., 200 g) to the PZT and measure the instantaneous change at the amplifier output at the time of loading.

    NOTE: If there is a drift in the output voltage and it is running out of the power supply range, temporally short the C1 terminals before applying the force. This will force the output voltage to zero line. The drift is due to the input bias current of the Op-Amp. TL084 has an FET input stage and the bias current is so small that the drift should be negligible.

13. Calculate the k coefficient of the PZT in units of Coulomb/Newton using the formulations given by:

$$\text{Charge Produced}\left(Q\right) = k \times \text{force}$$

$$\text{Output voltage}\left(Vo\right) = -Q/C = -k \times \text{force}/C \tag{5.1}$$

(Reminder: 1 kg = 1000 g = 9.81 N)

Use the initial voltage jump (Vo) in the output at the time of loading or unloading of the transducer. Note C is the value of the capacitor in the amplifier, excluding the intrinsic capacitor of the transducer. Why are we not considering the capacitance of the transducer here? (Hint: the input of the amplifier is virtual ground and thus at zero volts. The intrinsic capacitor of the transducer never gets charged up).

Mechanical Resonance

14. Now, let us measure the resonance frequency of the PZT transducer in an AC circuit. Connect the PZT in series to a 4.7 kΩ resistor and apply a sinusoidal waveform to the circuit from the signal generator with an amplitude of 10 Vpp or 20 Vpp (Figure 5.5).

15. Make sure that the PZT transducer is free hanging in the air (remove the tape). Any physical restriction on the transducer would change its resonance frequency.

16. Observe the voltage across the PZT using the oscilloscope while sweeping the input frequency from 100 Hz up to 20 kHz. Does the voltage make a dip at a certain frequency? Can you hear a buzz while going through audible frequencies of the human ear?

17. Make measurements of the PZT voltage at ten different points in steps of 1000 Hz up to 10 kHz and write down the values. Take extra points between 6–7 kHz at every 100 Hz (Figure 5.6). Plot the results in MATLAB as a function of frequency (Figure 5.7). The resonance frequency is where you see a dip in the voltage, which should be around 6.3 kHz. This means that the impedance of the transducer is minimum at the mechanical resonance frequency. Discuss why.

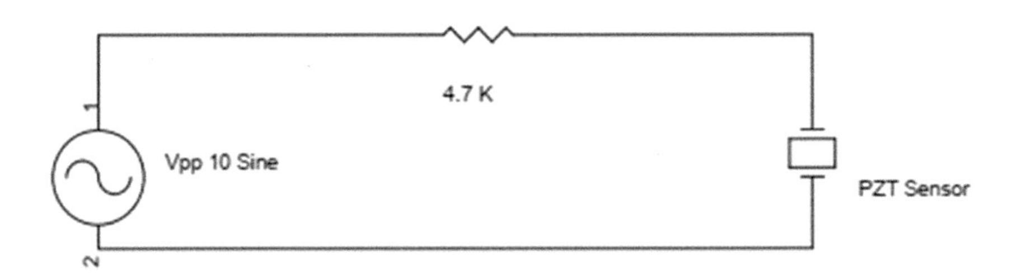

FIGURE 5.5 Circuit for measuring resonance. Plot the peak-to-peak voltage at the PZT side of the resistor.

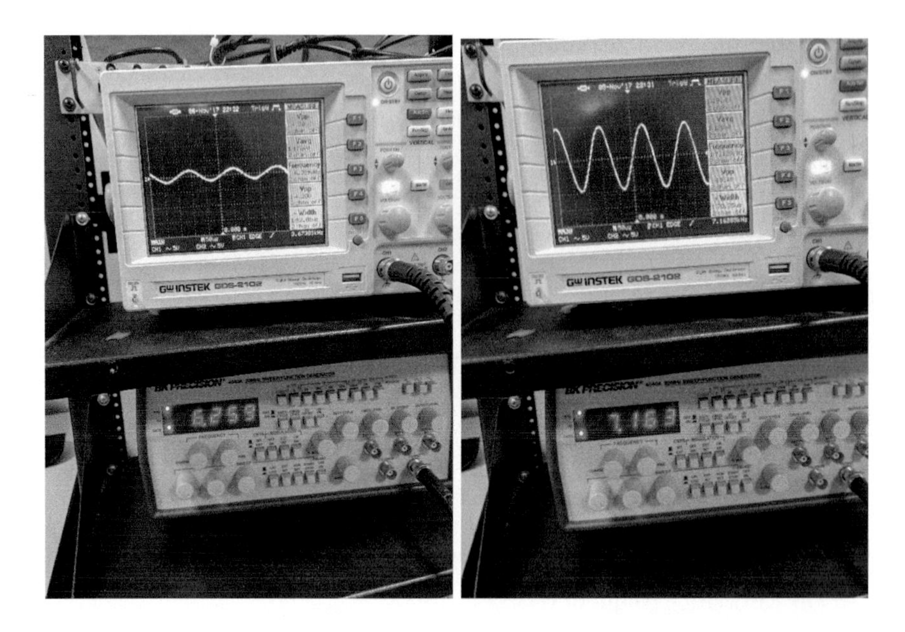

FIGURE 5.6 PZT voltage in Figure 5.5 at 6.259 kHz and 7.163 kHz showing the effect of mechanical resonance around 6.3 kHz.

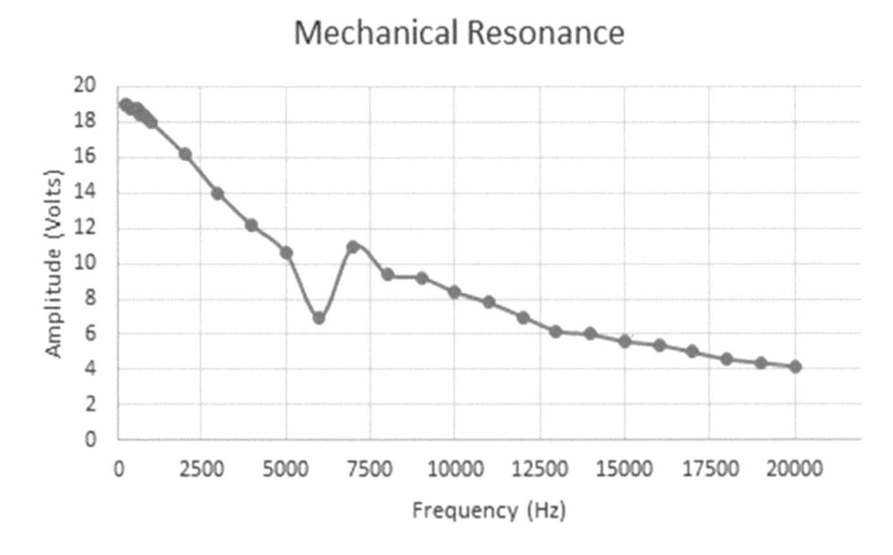

FIGURE 5.7 Voltage across the PZT transducer in Figure 5.5 as a function of frequency. The dip in the plot indicates the frequency of mechanical resonance.

S5.6 CIRCUIT TESTING AND TROUBLESHOOTING

1. Before turning the power on, carefully check if the voltage supply is applied to the Op-Amp with correct polarity. Using a voltmeter one lead connected to the ground, check the +/−9V and ground voltages at the power terminals of all chips. It is a very time saving practice to do this as the first step of troubleshooting in any circuit, analog or digital.
2. Another good practice is to measure the total current that the circuit is withdrawing from the power supply (most modern power supplies have a voltage and a current display). If the current is higher than a few tens of an mA, or the Op-Amp is becoming too warm (be careful not to burn your fingers) you may conclude a broken chip or a short circuit due to an incorrect connection.
3. Make sure that the PZT leads are firmly inserted into the protoboard and making good contact.

S5.7 DATA ANALYSIS AND REPORTING

In Results:

1. Remember to include answers to each one of the questions posed. Plot the waveforms using "volt" and "seconds" for the axes. Calculate the k coefficient and the total charge generated by the PZT due to the force applied using the output voltage waveform. The correct k values should be around 10–20 nC/N.

In Theory Section:

2. Background: Theory on PZT materials, mechanical resonance, and charge amplifier.

In Discussion:

3. Discuss the difference in the output waveform due to the presence/absence of the resistor and the capacitor value change.
4. Discuss potential biomedical applications of this project.

REFERENCES AND MATERIALS FOR FURTHER READING

References marked with an asterisk (*) are recommended to those interested in expanding on the content of this chapter.

1. *TL08xx Datasheet*, Texas Instruments, SLOS081I, May 2015, Dallas, TX.
2. *Murata catalog P37E-23*, page 5, January 28, 2010.
3. *Murata Specification Drawing*, JGB45-0419, April 20, 2005.
4. **Guide to the Measurement of Force*, The Institute of Measurement and Control, London, 2013. http://www.npl.co.uk/upload/pdf/forceguide.pdf.
5. Introduction to Piezoelectric Force Sensors, AllianTech, Gennervillers, France. http://www.alliantech.com/pdf/technique/force_piez_elec.pdf

STUDIO 6

Oscillometric Method for Measurement of Blood Pressure

S6.1 BACKGROUND

The sphygmomanometer is the instrument that physicians use to measure blood pressure with an arm cuff attached at the end. In 1896, Scipione Riva-Rocci invented an easy-to-use version of the mercury sphygmomanometer to measure brachial blood pressure [1]. The key element of this design was the use of a cuff that encircled the arm, whereas the previous designs had used rubber bulbs to manually compress the artery. Dr Nikolai Korotkov added the ability to detect systolic and diastolic pressures in 1905 with his discovery of Korotkoff sounds. Determining the systolic and diastolic blood pressures by listening to the Korotkoff sounds through a stethoscope inserted under the pressure cuff is called the auscultatory method. The pressurized cuff around the arm blocks the passage of blood through the main artery, either partially or completely, depending on the pressure. The first Korotkoff sound is heard when the cuff pressure is equal to the peak pressure during the ventricular contractions, which marks the systolic pressure. The blood flow becomes continuous, without interruption, when the cuff pressure is lower than the diastolic pressure during relaxation of the ventricles, at which point the Korotkoff sounds reduce substantially. However, the sound of cardiac pulsation is present in the background regardless of the blood flow, thus making the detection of systolic and diastolic pressures a somewhat subjective decision based on the changes in the intensity and texture of the sounds. In this studio we use the oscillometric method where the pressure oscillations inside the arm cuff generated by pulsatile blood flow are detected using a pressure sensor rather than listening to the sounds produced by it. The oscillometric method eliminates the need for a stethoscope and lends itself better to electronic measurement devices. Although somewhat less accurate, the oscillometric method is commonly used in automatic electronic blood pressure devices commercially available today.

S6.2 OVERVIEW OF THE EXPERIMENT

This laboratory exercise will utilize the Wheatstone Bridge introduced in Studio 1. The output of the Wheatstone Bridge pressure sensor is applied to an instrumentation amplifier to provide gain and then its single ended output to the Data Acquisition board to digitize the signal into the computer.

S6.3 LEARNING OBJECTIVES

The students will:

- Become familiar with pressure sensor operation.
- Use an Instrumentation Amplifier.
- Use the oscillometric method of blood pressure measurement.

S6.4 NOTES ON SAFETY

There is no health risk in this studio other than standard precautions that need to be observed in dealing with electronics.

S6.5 EQUIPMENT, TOOLS, ELECTRONIC COMPONENTS AND SOFTWARE

Electronic Components:

a. Pressure sensor (DPT-100, Deltran, see Figure 6.1).

b. Standard blood pressure cuff with a pressure dial.

c. Stethoscope.

d. Breadboard (protoboard).

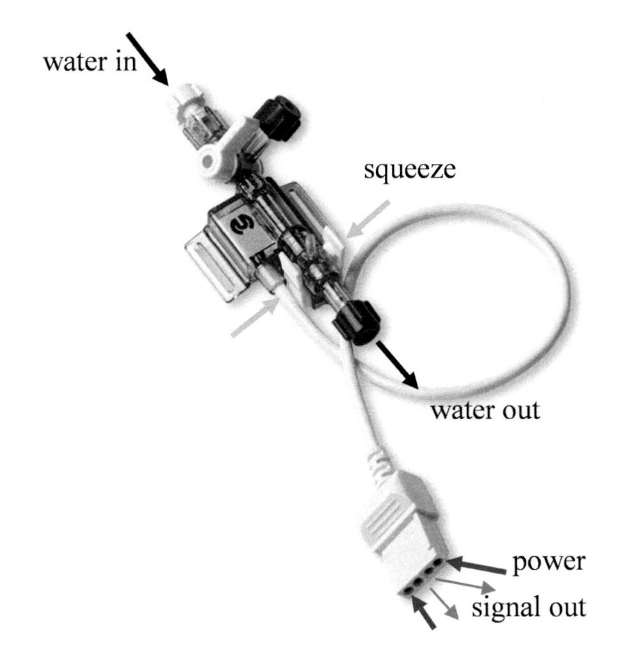

FIGURE 6.1 Disposable pressure sensor with a strain-gauge sensor in Wheatstone bridge configuration (DPT-100 at http://www.utahmed.com/deltran.html). The connector has four connections from the four corners of the Wheatstone bridge. Use the outer pair for power supply (red arrows) and the inner pair as the bridge output (blue arrows). Apply the pressure through the top end (open white cap) and use the other end (dark blue cap) for releasing pressure by squeezing the white plastic attachment (orange arrows). The stopcock points in the direction that the valve is closed to.

e. Instrumentation Amplifier INA126P.

f. A 10 cc or 20 cc syringe and tap water.

g. Ruler with centimeter divisions and a marker pen.

h. Various ¼ W 1% resistors and capacitors.

  2–100 Ω.

  1–1 KΩ.

  1–51 Ω.

  2–100 nF.

i. 2–2 pin header (optional).

j. 1–3 pin header (optional).

d. 1–100 Ω trimpot.

k. 2 BNC-to-micro clips cables.

Equipment:

- ±10 V power supply.
- Multimeter.
- A Data Acquisition Card (DAQ) installed into a computer.

Software:

- MATLAB®.

S6.6 CIRCUIT OPERATION

The circuit in Figure 6.2 shows the circuit to be built. The pressure sensor is made in the form of a Wheatstone bridge using four strain-gauges with a differential output. Instrumentation amplifiers as a topic were covered in Studio 2. The R1 sets the amplifier gain (IN126) to ×1000. The resistor network in the offset trim circuit is designed to produce a small positive or negative voltage around the ground potential by adjustment of a trimpot. This voltage is applied as a virtual ground point to the instrumentation amplifier (pin 5) to eliminate a small offset voltage from the amplifier's output. A few mV of an output offset may be due to imperfections in amplifier's manufacturing. Larger values of the output offset would be coming from the pressure sensor, which may have a small output at zero pressure. C1 and C2 are stabilizer capacitor to suppress the noise in the power supply and their value is not very critical.

S6.7 DETAILED EXPERIMENTAL PROCEDURE

1. The sensitivity of your pressure transducer DPT-100, Deltran) is ~5 µV/V/mmHg, but not exact. Calculate the voltage output for the amplifier in Figure 6.2 with the gain option 1000 selected, for pressure readings of 80 mmHg and 120 mmHg with this sensor.

2. Build the circuit in Figure 6.2. The connector has four connections from the four corners of the Wheatstone bridge. Use the outer pair for power supply (red arrows in Figure 6.1) and the inner pair as the bridge output (blue arrows). If a matching connector is not available

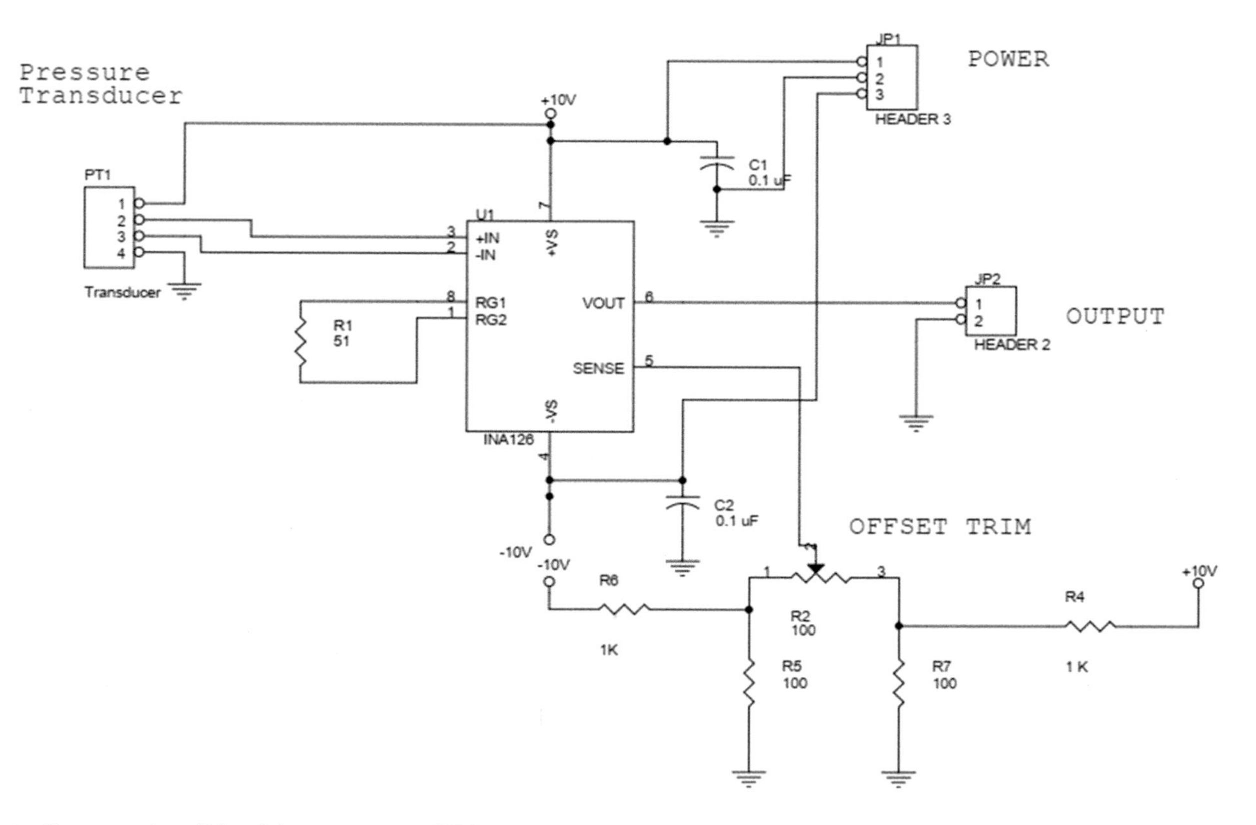

FIGURE 6.2 Pressure Amplifier Schematic using INA 126.

for connection, as a practical solution, you may cut the connector of the transducer with a cutter and insert the de-insulated end of the wires directly into the protoboard.

3. Open the white cap and the valve of the pressure sensor to room air (by turning the stopcock away from the white cap). Measure the circuit output with a DC voltmeter and adjust the offset trimpot to make sure that the output is zero for zero pressure.

4. Tape the Tygon® tubing that came with the pressure transducer on the wall vertically for its entire length (Figure 6.3). Attach the tubing to the port where the white cap was. Insert the 10 cc syringe filled with tap water to the port where the red cap was. Turn the stopcock 180° from the syringe. You will feel a stopper break while turning since it is not designed normally to turn in that direction. In this

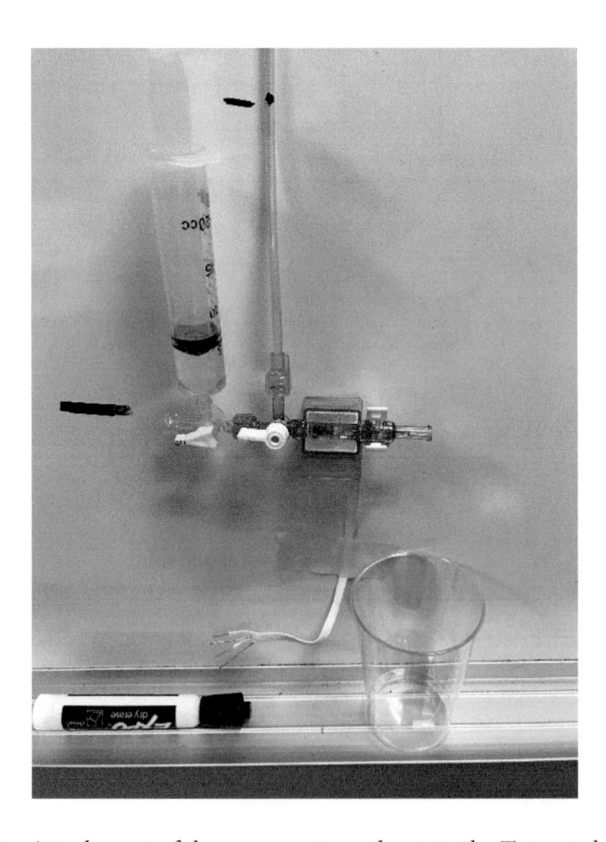

FIGURE 6.3 Attachment of the pressure transducer to the Tygon tubing and the syringe for calibration. The black mark indicates the zero point for the water column.

position, the pressure sensor will be open, both in the directions of the tubing and the syringe. Fill the tubing with tap water using the syringe. Once the water level reaches the desired level, the stopcock needs to be turned towards the syringe in order to make sure that the water pressure does not push the syringe piston back out. There should not be any air bubbles inside the tube or the pressure sensor. You may use the other end of the pressure sensor (dark blue cap side) to bleed out water and the air bubbles along with it by squeezing the white attachment (orange arrows). Mark the tubing at every 13.6 cm (13.6 cmH_2O = 10mmHg) as high as you can. The starting point should be at the same height as the transducer on the table. Note that the strain-gauge sensor is located where the black dot is inside the transparent casing. The height of this dot should be taken as the reference point for zero pressure height for the water column.

5. Calibrate the transducer by pushing water into the tubing and setting the water column at heights from 0–136 cm in steps of 13.6 cm while reading the output voltage with a DC voltmeter.

6. Make a chart in Excel or MATLAB showing output voltage as a function of pressure in mmHg as in Figure 6.4 and fit a straight line passing through the origin (interception zero). The slope of the line should give you the sensor's sensitivity in units of V/mmHg. Does

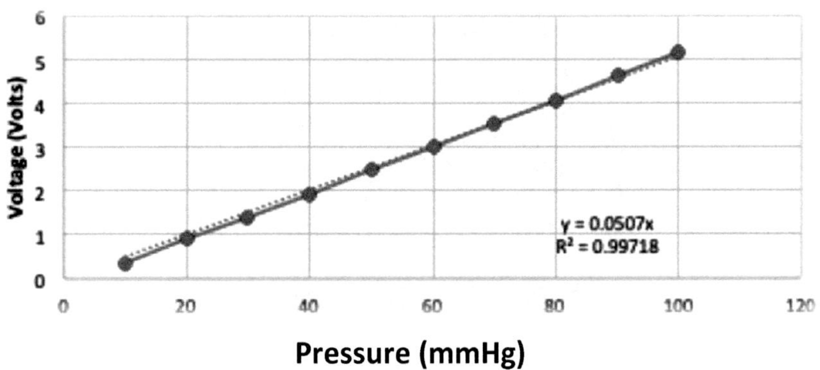

FIGURE 6.4 Typical calibration plot. The slope of 0.0507 V/mmHg is the calibration coefficient. Considering 10 V bridge exciting voltage and the gain of the Instrumentation amplifier (×1000), this calibration value corresponds to sensor sensitivity of 5.07 μV/V/mmHg.

the sensitivity of the pressure sensor given in the manufacturer's datasheet (5 μV/V/mmHg) match the measured value? Remember to account for the amplifier's gain of ×1000. Also, account for the Wheatstone bridge excitation voltage. For instance, the sensitivity becomes 50 μV/mmHg for an excitation voltage of 10 V. Each 13.6 cm of water column on the tubing corresponds to 10 mmHg.

7. Next, take the pressure cuff and connect it to the pressure transducer (as in Figure 6.5.) after removing the syringe, water and the Tygon tube. Tightly close the other end of the transducer with the red cap. Alternatively, you can use that port for connecting a mechanical sphygmomanometer that came with the arm cuff if you have to remove it from the cuff while attaching the cuff to the system. You will need the pressure measurements on the sphygmomanometer in order to measure the systolic and diastolic pressures using the auscultatory method and verify your pressure measurements made with the oscillatory method.

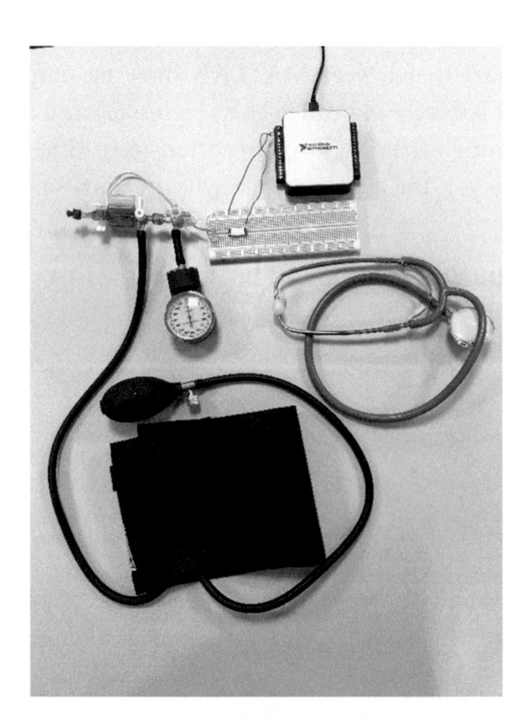

FIGURE 6.5 Connection of the transducer to the pressure cuff and the sphygmomanometer for measurements of blood pressure.

8. Run MATLAB on the computer and execute "*softscope*" command (or the "*analogInputRecorder*" in newer versions of MATLAB). Choose 1000 for sampling frequency and only one input channel.

9. Use the input range that is smallest possible to get the best amplitude resolution from the DAQ board. This is critical in order to minimize the quantization noise and record the small oscillatory components with maximum signal quality.

10. Connect the output of your circuit to the input channel of the DAQ board Interface in single-ended configuration, i.e., with respect to ground.

11. Put the cuff on a subject's arm as physicians do (watch the artery line marker on the cuff) and place the stethoscope under the distal edge of the cuff while listening to the sounds. Inflate and deflate the cuff while collecting data into MATLAB. Wait until you see full inflation/deflation curve on the screen, as in Figure 6.6. Note the air should be bled slowly to allow about 40–50 s for deflation in order to see multiple cardiac pulsations around systolic and diastolic pressures. Mark the systolic and diastolic pressures using the auscultation method while collecting data. After transferring data to MATLAB environment (explained below), collect data multiple times from the same subject and from multiple subjects for statistical analysis of data in your report by repeating this step.

12. Export the signal to MATLAB's workspace and plot as a function of time. You need to generate a time axis variable (recall that the sampling frequency is 1000 samples per second) for plotting the x-axis in units of seconds. Convert the vertical axis into units of mmHg using the calibration value you measured.

13. Discard the initial inflation region (up to time equals 12 s in Figure 6.6). The remaining signal should look like the one in Figure 6.7.

14. Using high pass and low pass filters, remove the baseline wandering and the high-frequency noise from the signal. (Use FFT of the signal to decide on the corner frequencies of the filters). The high-pass corner should be around 1 Hz and the low-pass around 10 Hz as the first approach. The extracted oscillatory signal will look like Figure 6.8.

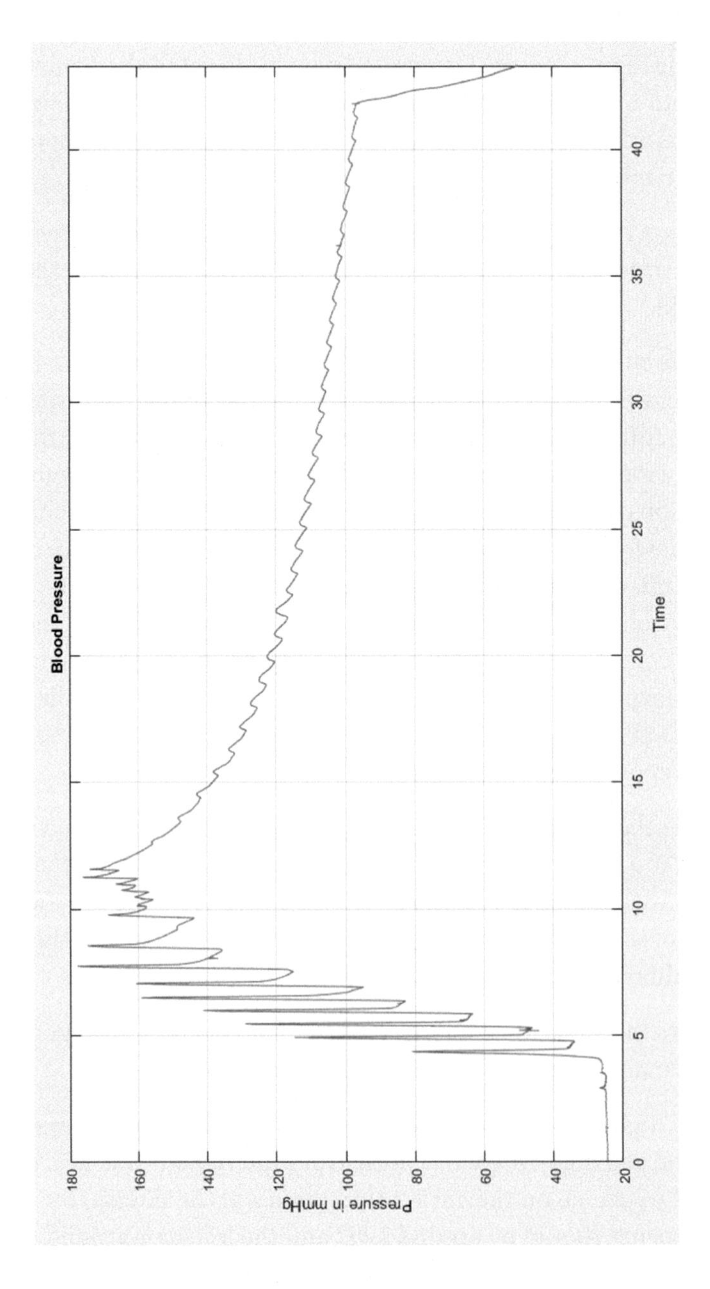

FIGURE 6.6 Typical inflation/deflation plot.

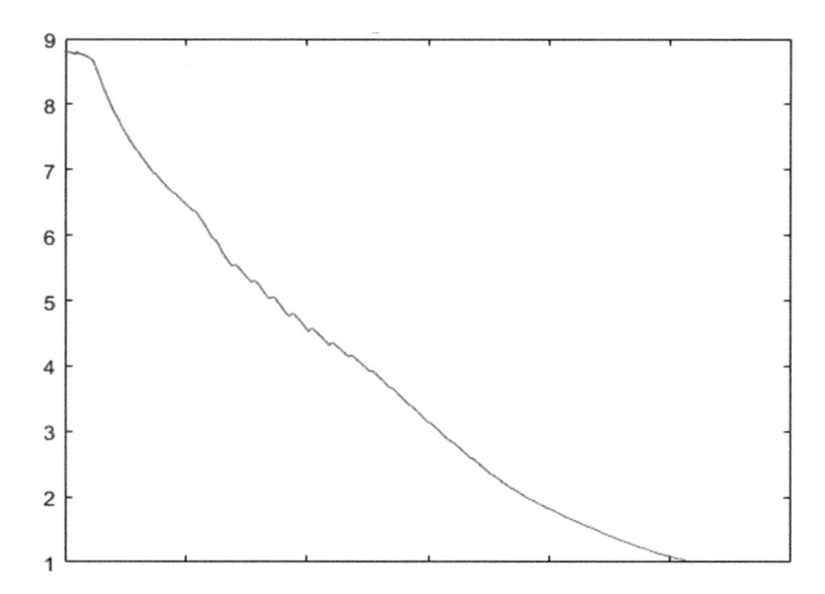

FIGURE 6.7 The pressure signal after removing the initial segment containing cuff inflation.

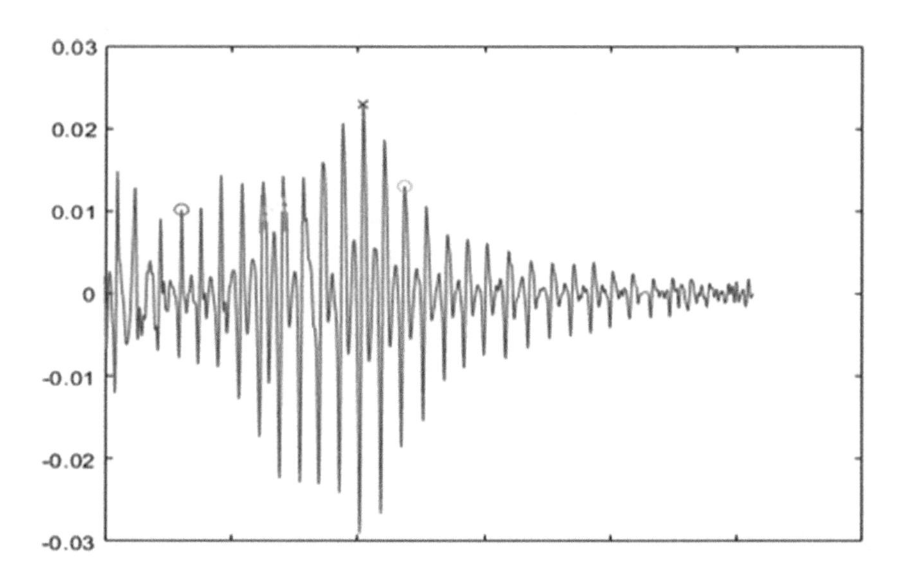

FIGURE 6.8 Extracted oscillations from the signal in Fig. 6.7 by band-pass filtering. The "x" marks the maximum amplitude of oscillations (the mean pressure), the "o" marks on the left and right indicate the oscillation amplitudes where the systolic and diastolic pressures are measured respectively.

15. In order to determine the mean pressure, find the point where the extracted oscillations have the highest amplitude (x mark in Figure 6.8). Then, find the corresponding time point in Figure 6.7 and read the pressure from the vertical axis. Following a similar method, determine the systolic pressure P_s by finding the first point that the oscillation amplitude reaches 0.55 P_m in Figure 6.8 and the blood pressure at the corresponding time point in Figure 6.7. Similarly determine the diastolic pressure, P_d, as the last point that the oscillation amplitude is reduced to 0.85 P_m. These coefficients (0.55 and 0.85) are estimates taken from a publication by Geddes et al. based on similar experiments in human subjects [2].

S6.8 CIRCUIT TESTING AND TROUBLESHOOTING

- Follow the general troubleshooting tips given in Studio 4 and at the end of Appendix I.
- Make sure the trimpot is connected properly. The middle point of the trimpot is also the middle terminal physically. If the output cannot be zeroed using the offset adjustment circuit, the output voltage from the sensor may be too large at zero pressure. Although it is normal to have a few mV of output at zero pressure, a larger value may indicate a broken sensor. Measure the bridge output directly with a voltmeter. Check the bridge excitation voltage and its connections to the power supply. As a final check before deciding a broken sensor, apply a small pressure or suction with the syringe attached to one of the ports to see if the output voltage changes at all. If this generates a smaller change than expected, the amplifier gain may be less than the assumed gain of ×1000.

NOTE: Using a small-caliber syringe, such as 1 cc, one can generate very large pressures, i.e., force per unit area, without realizing. This can easily overload the sensor and damage the strain-gauges inside. Large diameter syringes, 10 cc or 20 cc, are recommended. Refrain from applying large force to the syringe in any case.

S6.9 IMPORTANT TOPICS TO INCLUDE IN THE LAB REPORT

In Results:

1. The plot of output voltage vs. the calibrated pressure in mmHg (13 cmH_2O is equal to 10 mmHg).
2. The time plot of the pressure during inflation/deflation of the cuff (vertical axis is calibrated in mmHg).
3. Measurement of systolic and diastolic pressures from this plot.

In Theory Section:

4. Information on the pressure transducer used (www.utahmed.com/deltran.html - The model number is DPT-100).
5. Information on how the systolic and diastolic pressures are estimated with the oscillometric method (see the reference list below).

In Discussion:

6. The amount of measurement error you would have in mmHg units if you did not cancel the output offset with the offset adjustment circuit. Would this error be more significant at lower pressures or higher pressures?

REFERENCES AND MATERIAL FOR FURTHER READING

1. Riva-Rocci, Scipione. Un nuovo sfigmomanometro, *Gazzetta Medica di Torino* (1896) 47:1001–1017.
2. Geddes, L.A; Voelz, M; Combs, C: Reiner, D: and Babbs, Charles F.,. Characterization of the oscillometric method for measuring indirect blood pressure, *Ann. Biomed. Eng.* (1982) 10:271–280.
3. *INA126 Datasheet*, Texas Instruments, SBOS051D rev., January 2018, Dallas, TX.
4. *Deltran Datasheet*, http://www.utahmed.com/deltran.html.

STUDIO 7

Electronic Stethoscope

Heart Sounds

S7.1 BACKGROUND

Listening to heart sounds dates as far back as the 1700s when Jean Baptiste de Senac, a physician to King Louis XV of France and the author of the first known text on cardiology, used auscultation and percussion for diagnosis. The first listening aid or stethoscope was invented in 1816 by Laennec which he described in his Treatise on Mediated Auscultation in 1821. A stethoscope that amplifies auscultatory sounds is called a phonendoscope or electronic stethoscope. The four chambers of the heart have valves between the lower chambers, or ventricles, and the upper chambers, or atria. The tricuspid valve is on the pulmonary side that connects the right ventricle to the right atrium, and the mitral valve is on the other side that connects the left ventricle to the left atrium. In addition, there are two other valves, the pulmonary valve from the right atrium sending blood into the lungs, and the aortic valve sending blood to the entire body other than the lungs.

When a physician auscultates the heart, the physician is listening to the sounds from the valves. Abnormal sounds are called murmurs. Some murmurs are functional, i.e., they are harmless and do not affect cardiac function. Others are signs of disease, such as when the valves do not close properly due to stenosis or cardiomyopathy. Other causes of murmurs are heart defects, such as atrial septal defects (ASD), which is a hole in the wall between the two atria.

S7.2 OVERVIEW OF THE EXPERIMENT

This studio will implement an electronic stethoscope and observe heart sounds via auscultation, oscillography and spectrum analysis (FFT).

S7.3 LEARNING OBJECTIVES

The objectives of this studio are to:

- Learn the basics of auscultation.
- Understand the operation of an electret microphone.
- Learn data acquisition into a computer.
- Compute and plot the power spectrum of digitized signals.

S7.4 SAFETY NOTES

It is extremely important that the subject is not connected directly to the ground of any of the instruments used during this studio – e.g., oscilloscope, voltage supply, computer. In hospitals the clinical equipment has isolated grounds that prevent the passage of AC or DC current from such equipment to the patients. Isolated grounds are a safety measure to protect patients in case of faulty equipment. In the worst situation, a direct connection between the subject and the ground of a faulty apparatus may result in a fatality.

S7.5 EQUIPMENT, TOOLS, ELECTRONIC COMPONENTS, AND SOFTWARE

Additional information about the use of the items required in this studio can be found in the Introduction and in the Appendices.

Equipment:

- Breadboard.
- Power supply (+5 V).
- Multimeter.
- Oscilloscope.
- Data Acquisition Board (DAQ).
- Computer.

Tools:

- BNC-to-microclip cables.
- Banana plug to microclip cables.

Components:

a. Breadboard (protoboard).
b. Operational Amplifier LM833NG.
c. Various ¼ W resistors and capacitors (low leakage preferred).

    2–2.2 K.

    1–10 K.

    1–820 K.

    1–2.2 μF 16V.

    1–100 nF.

    1–15 μF 25V.
d. 1 Electret Microphone AOM-4546P-R PUI Audio.
e. 2-2 pin header.
f. Tubing 2” 5/16” ID latex rubber.

Software:

- MATLAB®.

S7.6 DETAILED EXPERIMENTAL PROCEDURE

1. As a preparation for this laboratory exercise, study the locations shown in Figure 7.1 over the chest and rib cage where the sounds of the four different heart valves can be listened to using a stethoscope.
2. Select two subjects in your study group and verify the best stethoscope positions by listening to the heart sounds for yourself.
3. Assemble the components on the breadboard as per the schematic in Figure 7.2.
4. Using JP1 connect the output of the amplifier to one of the inputs of the data acquisition board using the BNC to microclip cable and the

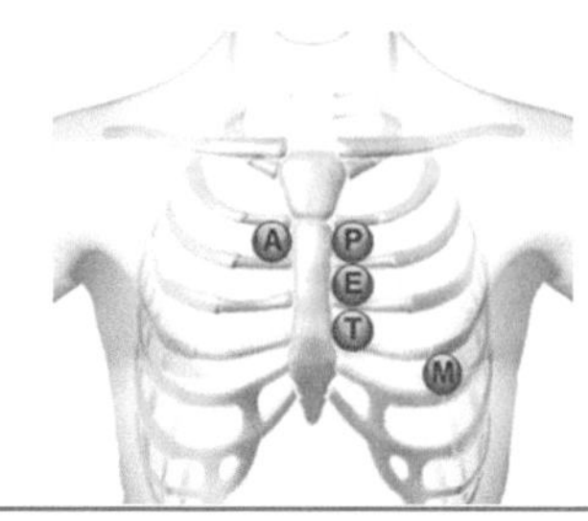

A	Aortic Valve Area	Second right intercostal space (ICS), right sternal border
P	Pulmonic Valve Area	Second left intercostal space (ICS), left sternal border
E	Erb's Point	Third left ICS, left sternal border
T	Tricuspid Valve Area	Fourth left ICS, left sternal border
M	Mitral Valve Area	Fifth ICS, left mid-clavicular line

FIGURE 7.1 Auscultation locations for each heart valve.

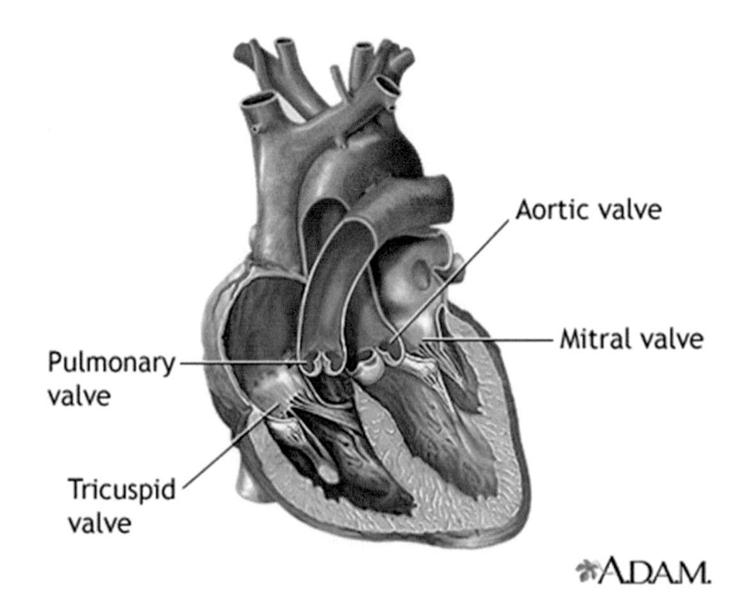

FIGURE 7.2 Cutaway cross section of a human heart. (https://medlineplus.gov/ency/article/003266.htm)

oscilloscope using a scope probe at the same time to visualize the signals. You may also connect the output of the amplifier to the input of a computer speaker (the kind that has an amplifier inside) in order to hear the sounds. Connect a power supply set to 12 V DC that will power the board + to JP2-1 and – to JP2-2. The power supply should be off when making the connections.

5. Attach the electret microphone (Figure 7.3) to a stethoscope by sliding a 2" length of 5/16" ID soft latex rubber tubing over the microphone. Then, remove the plastic tube from one of the earpieces on your stethoscope and insert it into the other end of the latex tube as shown in Figure 7.4.
6. Apply power to the microphone amplifier by turning on the power supply.
7. Run MATLAB on the computer and execute "softscope" command (or "analogInputRecorder" in newer version of MATLAB). Use a sampling rate of 1000 Hz.
8. Collect heart sounds from one of the subjects in the group by placing the stethoscope on his chest at those preselected points to detect different heart valve sounds. Collect each signal for about 30 seconds and save the data.

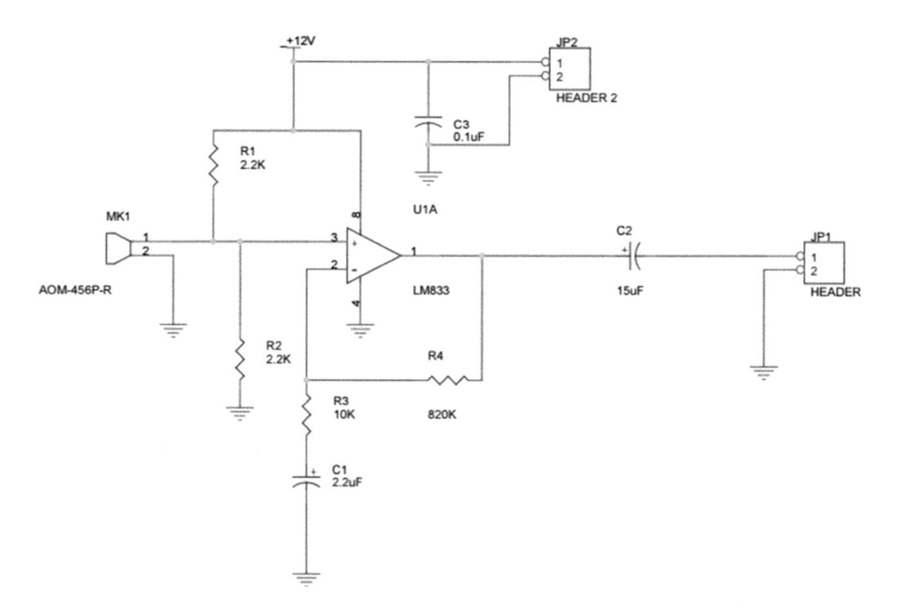

FIGURE 7.3 Electret microphone amplifier schematic.

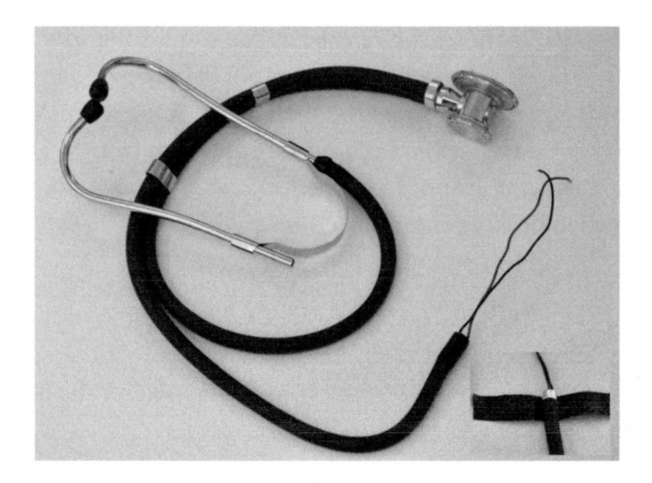

FIGURE 7.4 The electret microphone attached to the stethoscope tubing using an electric tape.

9. Plot the power spectrum of the heart valve sounds using *pmtm*, *pwelch* or other similar functions in MATLAB. Check the format using the help function in the command window.

10. Try to differentiate between different heart valves by using characteristic frequency peaks in the spectra.

11. Compare the spectra between two different subjects in the group. Determine the characteristic frequencies that are similar in both subjects.

12. Also, compare the temporal waveforms that are characteristics of different valves. Can you tell the difference when you listen to them through the speaker?

S7.7 CIRCUIT TESTING AND TROUBLESHOOTING

1. Follow the general troubleshooting steps recommended in previous studios.

2. The electret microphone has a built-in FET transistor for initial amplification of charges produced by the sound waves. Thus, the drain current (~1 mA) provided by R1 is essential for proper operation of the internal FET transistor. Confirm that the output of the microphone has a few volts DC bias, i.e., the transistor is not in saturation.

3. Note that this amplifier has a gain of 1 for DC signals but a gain of 82 (R4/R3) for AC signals. Thus, the Op-Amp output will have an offset that is around the same DC voltage measured at its input. The high AC gain is provided by including C1 in the feedback circuit. C2 is the coupling capacitor that blocks the DC voltage from appearing in the output voltage.

S7.8 DATA ANALYSIS AND REPORTING

In Results:

1. Include time and frequency plots of the recorded signals.
2. Mark and talk about characteristic frequencies.

In Theory Section:

1. Explain the source of heart sounds and their diagnostic value for various cardiac conditions.
2. Show the best stethoscope positions for listening to them on the chest.
3. Provide information about their characteristic frequencies and temporal waveforms from the literature.
4. Briefly describe how various stages of the microphone amplifier circuit works.

In Discussion:

1. Discuss your data in light of the information you found on heart sounds from literature and publications online.

REFERENCES AND MATERIALS FOR FURTHER READING

References marked with an asterisk (*) are recommended to those interested in expanding on the content of this chapter.

1. *LM833 Datasheet*, ON Semiconductor, LM833D, September 2011, Rev. 6.
2. *AOM-4546P-R Microphone Datasheet*, PUI Audio, Rev. A.

STUDIO 8

Transmission Photoplethysmograph

Fingertip Optical Heart Rate Monitor

S8.1 BACKGROUND

For centuries humankind has been aware that the frequency of the heartbeat is not constant, but it varies with multiple factors, yielding what is known as *heart rate variability* (HRV, see Studio 12). Galen of Pergamon (AD 131–200), a physician in ancient Rome, was the first to report that physical exercise causes changes in the heart rate [1]. Ibn al-Nafis (1213–1288), who first described the pulmonary circulation of the blood, observed that the arteries contract when the heart expands and expand when the heart contracts [7]. Much later, in the nineteenth century, Carl Ludwig showed that the heart rate increases during inspiration and decreases during expiration; this dependence of the heart rate upon respiration is known as *Respiratory Sinus Arrhythmia* or RSA. Studies by Hon in the 1960s reported that the heart rate of the fetus increases prior to episodes of fetal distress that can lead to death [2]. The simplicity, low cost and portability of the photoplethysmograph (PPG) [7] has made it a standard of care in many clinical settings including operating rooms, intensive care units and delivery rooms. Most recently, PPGs have been developed that connect directly to cell phones and record an individual's heart rate far from a clinical setting and even while they are exercising [5].

Photoplethysmographs (after the Greek words: *photo*, light; *plethysmo*, increase; *graph*, to record) are able to detect the increase in tissue volume due to the accumulation of blood during the systolic phase of the cardiac cycle. Such an increase in volume is more evident in tissues where the capillary bed is denser as in fingers, toes, ear lobes, or nostrils. The PPG consists of a *light source* that injects light into the tissue and a *light detector* that collects the light returned by the tissue. In the configuration used here, light is detected after crossing the whole tissue – e.g., a finger (Figure 8.1A). This configuration is known as *transmission photoplethysmograph*. Conversely, when the detector collects the light reflected by the tissue – i.e., the light that does not cross the tissue – the configuration is called *reflectance photoplethysmograph*. During systole blood accumulates in the very narrow capillaries and it becomes more difficult for light to cross the tissue through the higher volume of blood. As a result, the intensity of transmitted light decreases in every systole and returns to a basal level in every diastole (Figure 8.1B). The changes in transmitted light (often called the AC component of the PPG signal) are associated to oxygenated (arterial) blood whereas the basal level (DC component) is mostly linked to light absorption by skin pigments, bloodless tissue and deoxygenated (venous) blood [6].

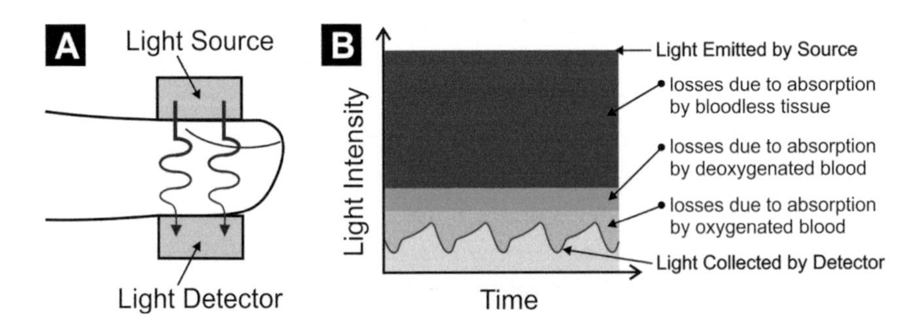

FIGURE 8.1 (A) Schematic of the experimental setup for the finger photoplethysmograph showing how the finger is located in the optical path between the light source (LED) and the light detector (phototransistor). (B) Decrease in the intensity of light (*light emitted by the diode*), as it crosses the finger of a subject, due to absorption by the tissue (skin and flesh) and the bone. The lower curve displays the intensity of light that reaches the photodetector of the photoplethysmograph, which replicates the pulsatile wave of the cardiac cycle.

The transmission of light through a material is quantified through transmittance (T), which is defined as the ratio of exiting (I) to incident (I_0) light intensities:

$$T = I / I_0 \tag{8.1}$$

Alternatively, the difference between injected and transmitted light intensity can be expressed by absorbance (A), which is the reciprocal to the transmittance:

$$A = 1 / T = I_0 / I \tag{8.2}$$

Absorbance can be related to the optical properties of the material and the thickness of the material crossed through the Beer-Lambert law:

$$A = 10^{\varepsilon LC} \tag{8.3}$$

With ε, L and C standing for the *extinction coefficient* (also called *molar absorptivity*), the *optical path length* and the *concentration* of the optically absorbent species in the material, respectively. The higher the extinction coefficient the higher the absorbance and the lower the intensity of transmitted light. It is important to note that the product of the extinction coefficient times the concentration is commonly known as the *absorption coefficient* (α), which yields an alternative form of the Beer-Lambert law:

$$A = 10^{\alpha L} \tag{8.4}$$

The extinction coefficients have been experimentally determined for the different forms of hemoglobin, which is the molecule responsible for most of the transport of oxygen and oxygen dioxide in blood (Figure 8.2). The form of hemoglobin bound to oxygen is called *oxyhemoglobin* (HbO_2), the unbound hemoglobin is *deoxyhemoglobin* (Hb) and hemoglobin bound to carbon oxide is *carboxyhemoglobin* (HbCO), a molecule commonly found in the blood of smokers. The extinction coefficients of the hemoglobin forms depend strongly on the wavelength of the incident light. Three wavelengths are relevant to the operation of photoplethysmographs. (i) At 660 nm the difference between the extinction coefficients of Hb and HbO_2 is maximum, with oxyhemoglobin exhibiting an extinction coefficient about ten times smaller than that of deoxyhemoglobin. As a result, readings at 660 nm are

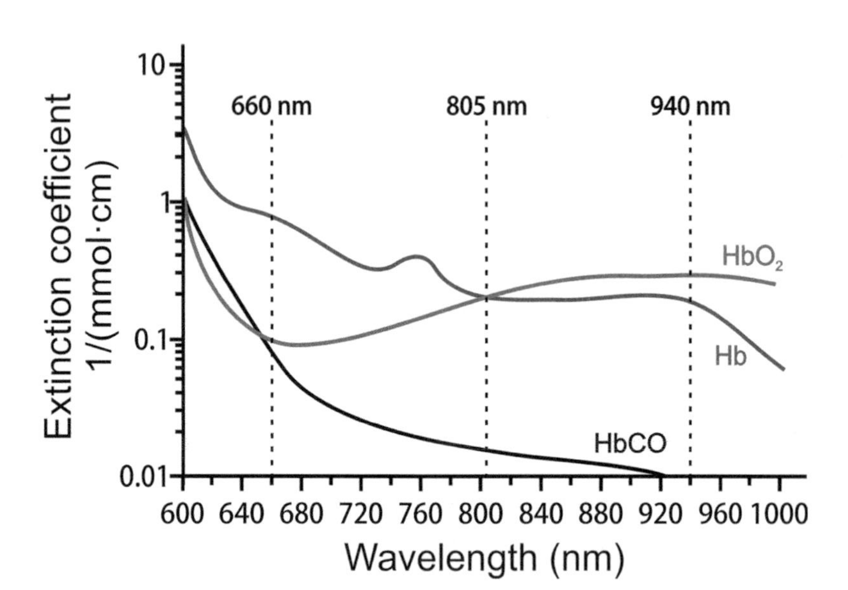

FIGURE 8.2 Extinction coefficients (or molar absorptivities, in units of $mmol^{-1} \cdot cm^{-1}$) for the three most common forms of hemoglobin species (oxyhemoglobin, deoxyhemoglobin and carboxyhemoglobin) at the wavelengths of interest in photoplethysmography. (Adapted from [6].)

mostly dependent on the volume of arterial (oxygenated) blood present. (ii) The extinction coefficient of HbO_2 becomes larger than that of Hb at infrared (IR) wavelengths. Measurements in the IR region of the spectrum for PPGs are typically performed at 940 nm, where the extinction coefficient of HbO2 presents a local maximum. (iii) The extinction coefficients of Hb and HbO_2 are identical at the *isosbestic wavelength* of 805 nm and therefore optical measurements at this wavelength are independent of the level of oxygenation of the blood. Measurements at the isosbestic wavelength can be used to normalize the readings at other wavelengths and compensate for the effects (e.g., varying scattering) that may result in differences in hematocrit content from patient to patient. In this studio we will perform measurements at either 660 nm (red) or 940 nm (infrared). Despite that, we will not take readings at the two wavelengths simultaneously, some commercial apparatuses do so and are able to extract an approximate value of the oxygen saturation in blood from the comparison between the readings at the two wavelengths [6].

Two light-emitting diodes (LEDs, one for each wavelength) will act as the light source in this studio. The first section of our plethysmograph (left in Figure 8.3) will power the LED with a voltage source (V_{DC}) and

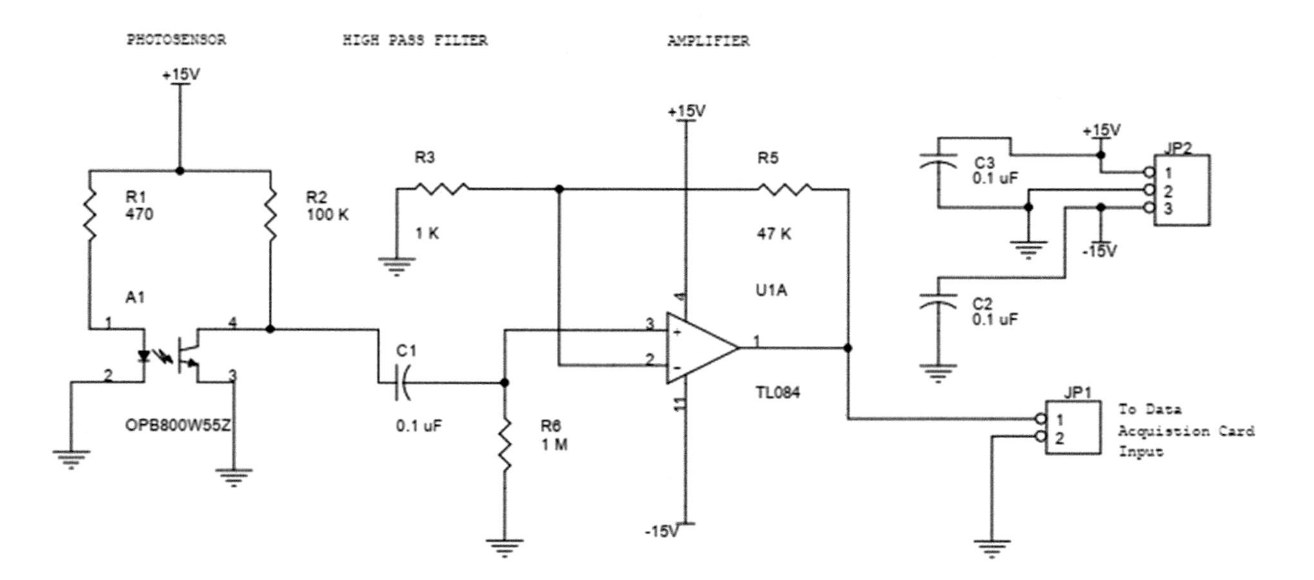

FIGURE 8.3 Schematic of the circuit for the finger transmission photoplethysmograph. The output of the circuit will be connected to a Data-Acquisition Board (DAQ) in order to record the measurements directly into a computer.

a biasing resistor (R_1) that will limit the current allowed to flow through the LED, protecting the LED from burning. A phototransistor will be used as the light detector that will collect the light emitted by the LED after it traverses the finger of a subject. The voltage at the collector of the phototransistor (v_{COL}) will be related to the intensity of light detected. The circuit includes a resistor (R_2) that protects the phototransistor from excessive current. The block identified as PHOTOSENSOR in Figure 8.3 comprises the light source and detector and the components required to power them. The next block in the *signal pathway* – i.e., going from input to output, or simply left to right in Figure 8.3 – is a HIGH-PASS FILTER. The filter consists of a resistor (R_6) and a capacitor (C_1) and its cutoff frequency can be expressed as:

$$f_C = \frac{1}{2\pi R_6 C_1} \tag{8.5}$$

As discussed, the intensity of transmitted light decreases in every systole and returns to a basal level in every diastole (Figure 8.1B). The basal level of light (DC signal component) collected by the phototransistor will vary with several factors such as the skin pigmentation, the thickness of the finger, or the ratio of fat and bone in the finger [6]. Because the goal of a plethysmograph is to detect the heartbeats, our circuit (Figure 8.3) focuses on detecting the AC component of the transmitted light that varies with the cardiac cycle. The built-in high-pass filter and its capacitor allow high frequencies pass while blocking the DC signals and frequencies lower than f_C. The next block in the circuit is an AMPLIFIER that magnifies the signal (v_{OA}) before sending it to the Data Acquisition System. The transfer function of the amplifier is:

$$\frac{v_O}{v_{OA}} = 1 + \frac{R_5}{R_3} \tag{8.6}$$

The final step of the studio is processing the experimental data using MATLAB.

S8.2 OVERVIEW OF THE EXPERIMENT

This studio will implement a transmission PPG consisting of a LED and a phototransistor fixed on opposite sides of the index finger of a subject (Figure 8.1). Two different LEDs will generate light at 660 nm first and later at 940 nm, which will be detected by the phototransistor after traversing the finger. A circuit will be built on the protoboard to block the DC components and amplify the optical signal collected by the phototransistor. The signal of the heartbeat will be ultimately sent to a computer, where a script in MATLAB will automatically process the experimental data and determine the instantaneous and average heart rates.

S8.3 LEARNING OBJECTIVES

The objectives of this studio are to:

1. Learn the fundamentals of pulse oximeters.
2. Experiment with optical sensors.
3. Become familiar with dependence of the optical absorbance of hemoglobin upon its degree of oxygenation.
4. Understand and implement circuits for detection of cardiac pulsation.
5. Develop MATLAB® scripts for cardiac cycle analysis.

S8.4 SAFETY NOTES

It is extremely important that the subject is not connected directly to the ground of any of the instruments used during this studio – e.g., oscilloscope, voltage supply, computer. In hospitals the clinical equipment has isolated grounds that prevent the passage of AC or DC current from such equipment to the patients. Isolated grounds are a safety measure to protect patients in case of faulty equipment. In the worst situation, a direct connection between the subject and the ground of a faulty apparatus may result in a fatality. All electric equipment in the laboratory should be tested for safety prior to use. This is normally the responsibility of the host institution.

High-intensity LEDs may present a risk of retinal damage if brought near the eye. Students should be cautious not to look into the LEDs directly at a close distance – i.e., less than 20 cm. Additionally, the resistor that is connected to the LED (R_1 = 470 Ω, ½ W, Figure 8.3) should be handled with care as it may become hot to the touch.

S8.5 EQUIPMENT, TOOLS, ELECTRONIC COMPONENTS AND SOFTWARE

Additional information about the use of the items required in this studio can be found in the Introduction and in the Appendices.

Equipment:

- Breadboard.
- Double voltage supply (±15 V).
- Multimeter.
- Oscilloscope.
- Data Acquisition Board (DAQ).
- Computer.

Tools:

- Paperclip.
- BNC-to-microclip cables.
- Jumper wires for the breadboard.

Electronic Components (remember to download the available datasheets):

- Photointerrupter (OPB800W55Z, Optek Technology).
- Operational Amplifier (for example, TL084 or LM324).
- Resistors (¼ W): 2 × 1 MΩ, 1 kΩ and 100 kΩ.
- Resistors (½ W): 470 Ω.
- Capacitor: 100 nF.

Software:

- MATLAB®.

S8.6 DETAILED EXPERIMENTAL PROCEDURE

Build the circuit shown in Figure 8.3 on the breadboard using the specified components.

TIP: The Introduction provides a summary of best practices for using the breadboard.

ATTENTION: The LED needs to be connected so that the voltage at the anode (indicated with a triangle in the symbol of the LED) is higher than that at the cathode (indicated with a flat bar in the symbol of the LED), which results in a current flowing from the anode to the cathode.

1. Insert the index finger of a subject between the LED and the phototransistor as shown in Figure 8.4.

   ATTENTION: The phototransistor detects light at wavelengths other than that of the LED. As a result, environmental lights are detected and become noise added to the measurements of the PPG. To minimize the environmental noise, cover the LED-finger-phototransistor with an opaque material such as a cardboard box or a black paper.

2. Using BNC-to-microclip cables connect the output of the phototransistor (v_{COL}) to the oscilloscope. Power the circuit and observe on the oscilloscope screen a *positive-going* pulse – i.e., the tip of the pulse is higher than the rest of the pulse – for each heartbeat of the subject (Figure 8.5).

   TIP: The circuit should be tested following the signal pathway (from the input to the output), stage-by-stage, as it is always the best practice with any electronic circuit. When measuring voltages with a voltmeter, connect one probe to the point of interest in the circuit– e.g., a wire or the pin of a chip – and the other one to the ground of the circuit. By measuring each voltage with reference to the ground of the circuit one can bypass possible bad connections along the signal pathway.

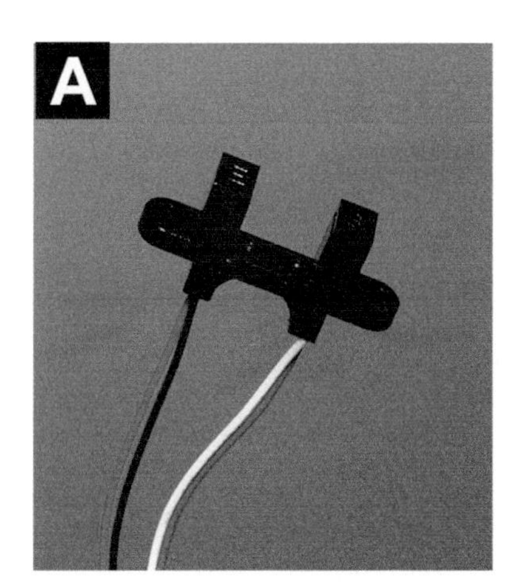

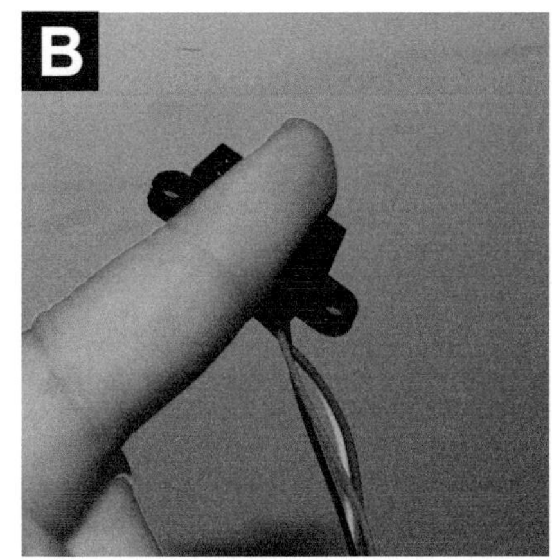

FIGURE 8.4 Photographs of (A) a commercial opto-interrupter (OPB800W) that can be easily transformed into the front end of a photoplethysmograph (B).

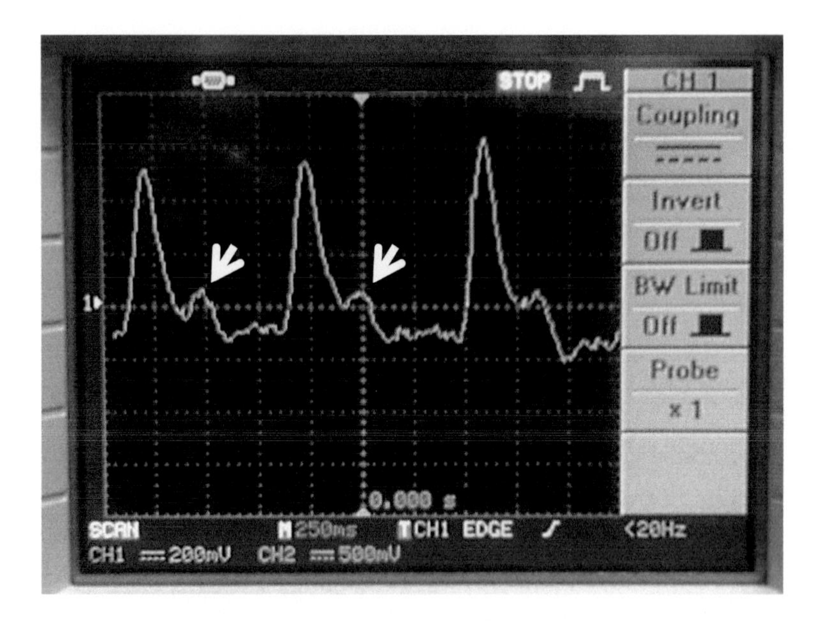

FIGURE 8.5 Example of signal collected with the photoplethysmograph in Figure 8.3, as displayed on the screen of an oscilloscope. The threshold for detection should be chosen around 0.3 V in this example to avoid the small peak (white arrows) while staying below the large peaks. CH1 is set to 200 mV/div.

ATTENTION: Before turning the power on, confirm that the supply voltage applied to the operational amplifier (Op-Amp) has the correct polarity. Modern Op-Amps are able to tolerate many types of circuital errors such as short-circuited outputs, excessive voltage applied to the inputs, etc. However, when the supply voltage is applied to the Op-Amp with the reverse polarity the chip burns in a fraction of a second. Notably, if the Op-Amp burns quickly, the plastic package covering it may not burn or even heat up.

TIP: If the 660 nm LED does not emit visible light, confirm that the LED is correctly polarized with a higher voltage at the anode than at the cathode.

TIP: If the LED is emitting light but pulses are not visible on the screen of the oscilloscope, confirm that the DC voltage at the collector (v_{COL}) is approximately halfway between the supply voltage $V_{S,}$ (+15 V) and 0 V. This PPG requires the phototransistor to operate linearly, providing an output proportional to the collected light. The phototransistor operates linearly when the DC voltage at the collector is far from V_S (when the transistor is in cutoff) and from 0 V (when the transistor becomes saturated). If the signal in the oscilloscope is too large and only high-low transitions are visible, ensure that the finger is correctly positioned between the LED and the phototransistor, which will avoid the phototransistor receiving too much light and going into saturation (v_{COL} <0.2 V).

3. Connect the output of the phototransistor (v_{COL}) and the operational amplifier (v_O) to each of the two channels of the oscilloscope. Verify that the signal from the phototransistor is amplified by the circuit as expected: with a gain of 100 for the resistors given in step 1.

TIP: The Introduction chapter provides a summary of best practices for using the oscilloscope as well as other equipment in this lab and the breadboard.

TIP: If the output signal of the amplifier differs from the one expected, (i) ensure the Op-Amp chip and all its pins are correctly pushed into the breadboard – sometimes a few pins may have been curled underneath without going into the corresponding holes of the breadboard, and (ii) verify the positive and negative supply voltages of the Op-Amp are as indicated in the datasheet.

ATTENTION: The signal at the output of the amplifier should have the same polarity as that at the collector of the phototransistor – i.e., positive-going pulses – because this circuit consists of a non-inverting amplifier. The output of the amplifier, however, should not show the DC offset present at the output of the phototransistor, which is blocked by the capacitor at the input of the amplifier (Figure 8.5).

TIP: Confirm that the signal at the non-inverting input of the Op-Amp is identical (showing no DC offset) to that at the inverting input. If the signal at the non-inverting input appears distorted or different from the signal at the collector of the phototransistor, verify that the value of the capacitor is actually 0.1 μF. The capacitor section in the Appendices describes the different industrial codes used to express the value of a capacitor.

4. After validating the circuit with the oscilloscope, apply the output signal to one of the analog inputs of the Data Acquisition Board (DAQ) using the appropriate cable: a BNC-to-microclip cable if you are using a DAQ Interface with BNC inputs.

5. Open the m-file named "Acquire_HeartRate_legacyversion.m" or "Acquire_HeartRate_sessionbased.m" on the website (http://www.routledge.com/9781466504660), depending on whether the version of the MATLAB software you have supports session-based data acquisition.

6. Set the threshold values somewhere in the middle according to the observed signal on the oscilloscope, such that the rising and falling edges are always detected despite variations in the signal amplitude. Run the MATLAB code and observe the heart rate displayed on the screen, which is stored in the variable "Heart_Rate." Notice that this program also outputs the heart rate as a voltage through one of the analog output ports of the DAQ Board, in case it needs to be used as a control signal elsewhere. Discuss what this might be useful for in your report.

7. Save the instantaneous heart rates in an array in MATLAB by modifying the code outside the while loop. Record for a few minutes the PPG signal from a subject in your group and save the collected data on the hard disk of the computer or your own flash drive.

   ATTENTION: Light with a wavelength of 940 nm is not visible. Non-visible light, however, can still cause retinal damage if brought near the eye so students should avoid looking into the LED directly at a close distance – i.e., less than 20 cm.

8. Modify the m-file so that the average of the last n instantaneous heart rate measurements is calculated and plotted on the computer screen as a *running average* of the heart rate. Typical values for n could be 4 – 8. Notice this produces much more stable heart rate values then the instantaneous rates.

9. Collect additional data from another subject in the group if time allows.

S8.7 DATA ANALYSIS AND REPORTING

The signal will be collected continuously in a loop while the *Acquire_HeartRate* code is running. There is sufficient time (10 ms) between consecutive samples to process the signal and calculate the heart rate when the cardiac cycle is complete. After initialization steps (read the comments next to each MATLAB line for explanations), the algorithm starts looking for a signal value above a certain threshold. This threshold value (Thresh) should be adjusted according to the signal amplitude observed on the oscilloscope screen so that it is set properly on the rising side of the signal. Once the rising edge is detected, it moves to the next loop looking for the falling edge. Detection of a falling signal edge indicates the end of one beat, but the next cycle does not begin until another rising edge is detected. Only then, the code calculates the time interval between the two rising edges that demarcate the start of two consecutive cardiac cycles. The inverse of the inter-beat interval (IBI) is the instantaneous heart rate. To express this value in beats-per-minute (BPM) it should be multiplied by 60.

The MATLAB code also sends the instantaneous heart rate to one of the analog output terminals of the DAQ Board (A0) as a voltage. Since DAQ board cannot generate 70–80 V, the heart rate is scaled down by a factor of 100 before it is copied to the output. This output voltage can be used to display the heart rate on a voltmeter or to trigger other events (alarms etc.) if it is above a certain value. The voltage output is included here as an example of how DAQ Board can be utilized to send signals out using MATLAB.

Important Topics to Include in Lab Report

a. Discuss any methodological similarities that are learned in this laboratory and the journal article on *Vital Signs: Heart Rate* by Michael R. Neuman [2].

b. Discuss the change of absorption of light as a function of wavelength by oxyhemoglobin and deoxyhemoglobin, and the isosbestic wavelength.

c. Discuss the sensitivity of this device to positioning of the finger in the probe and potential improvements to the system.

d. Describe any troubleshooting you had to do and how you identified the problem.

S8.8 PRE-LAB QUESTIONS

1. How many decades are there between the cutoff frequency of the circuit in Figure 8.3 and the frequency of the power line (also known as

utility frequency)? Hint: a first-order filter does not have a very sharp transition band and thus a high-pass filter with a corner frequency that is much higher than 60 Hz may still pass signals at the power line frequency.

2. Will the DC signal at the input of the operational amplifier increase or decrease if C1 is exchanged by a 10 nF capacitor?

S8.9 POST-LAB QUESTIONS

1. What is the physiological phenomenon behind the 'kink' observed in the signal waveform (see the arrows in Figure 8.5)?
2. How can the sensitivity to ambient light be reduced by changing the optical bandwidth of the photodiode?
3. Can the signal-to-noise ratio be improved by increasing the LED light intensity?
4. Can the same measurements be made at the isosbestic wavelength (805 nm)? Why not?
5. Can the signal amplitude be increased by using two LEDs, one red and the other infrared at the same time, and taking the difference between the two photodiode outputs?
6. What are the signal polarities (positive-going or negative-going) at each wavelength?

REFERENCES AND MATERIALS FOR FURTHER READING

References marked with an asterisk (*) are recommended to those interested in expanding on the content of this chapter.

1. Galen, *The Pulse for Beginners*, in: P.N. Singer (translator), Oxford University Press, New York, 1997, pp. 332.
2. Ludwig, C. Beitrage zur Kenntnniss des Einflusses der Resprirations bewegungen auf den Blutlauf im Aortensysteme, *Arch. Anat. Physiol.* (1847) 13:242–302.
3. Hon, E.H.; Lee, S.T. Electronic evaluations of fetal heart rate patterns preceding fetal death, further observations, *Am. J. Obstet. Gynecol.* (1965) 87:814–826.
4. *Neuman, M.R. Vital signs: heart rate, *IEEE Pulse* (2010) 1(3):51–55.

5. Furchgott, R. The Argus app can help to keep you fit, *New York Times*, July 23, 2013, http://www.nytimes.com/2013/07/25/technology/personaltech/the-argus-app-can-help-to-keep-you-fit.html.
6. Robert A. Peura, *Chemical biosensors, Chapter 10 in *Medical Instrumentation: Application and Design*, ed. J.G. Webster, 4th ed. John Wiley & Sons Inc., 2009.
7. Fancy, D. Nayhan. *Ibn Al-Nafīs and Pulmonary Transit*. Qatar National Library. Retrieved April 22, 2015.

Measurement of Hand Tremor Forces with Strain-Gauge Force Transducer

S9.1 BACKGROUND

A tremor is an involuntary, rhythmic muscle contraction leading to shaking movements in one or more parts of the body. It is a common movement disorder that most often affects the hands but can also occur in the arms, head, vocal cords, torso and legs. Tremor may be intermittent or constant. It can occur on its own or it can happen as a result of another neurological disorder. These oscillations that occur due to muscle contractions are believed to be a functional component of the neurological feedback loop control mechanisms. Small amplitude tremors are normal for healthy people.

Tremors can be classified into two main categories, resting tremor and action tremor. A resting tremor occurs when the muscle is relaxed, such as when the hands are resting on the lap. This type of tremor is often seen in people with Parkinson's disease and is called a "pillrolling" tremor because the circular finger and hand movements resemble rolling of small objects or pills in the hand. The second main category is the action tremor. This occurs with the voluntary movement of a muscle. There are several sub classifications of action tremor, for example, postural, kinetic, intention, task-specific and isometric tremor. The intention tremor is produced with purposeful movement toward a target, such as lifting a finger to touch the nose or moving a leg to kick a ball.

S9.2 OVERVIEW OF THE EXPERIMENT

The aim of this studio is to observe and analyze muscle tremors. An instrumentation amplifier (IA) is assembled and used to amplify the output of a force transducer. The tremor in the subject's index finger will be analyzed using the power spectra functions in MATLAB®. As an extra activity in this studio, one can also attach an accelerometer to the finger as a secondary source of tremor information. The easy-to-use three-axis accelerometer ADXL335 from spurkfun.com is recommended.

S9.3 LEARNING OBJECTIVES

The objectives of this studio are to:

- Understand the operation of the force transducer amplifier system.
- Become aware of problems during acquisition of noisy signals into the computer.
- Compute and plot the power spectrum of force signals.
- Become familiar with tremor as a physiological phenomenon.

S9.4 SAFETY NOTES

It is extremely important that the subject is not connected directly to the ground of any of the instruments used during this studio – e.g., oscilloscope, voltage supply, computer. In hospitals the clinical equipment has isolated grounds that prevent the passage of AC or DC current from such equipment to the patients. Isolated grounds are a safety measure to protect patients in case of faulty equipment. In the worst situation, a direct connection between the subject and the ground of a faulty apparatus may result in a fatality.

S9.5 EQUIPMENT, TOOLS, ELECTRONIC COMPONENTS AND SOFTWARE

Additional information about the use of the items required in this studio can be found in the Introduction and in the Appendices.

Equipment:

- Breadboard.
- Power supply (±5 V).
- Multimeter.
- Oscilloscope.
- Computer with a Data Acquisition Board (DAQ).

Tools:

- BNC-to-microclip cables.
- Banana plug to microclip cables.

Components:

a. Breadboard (protoboard).
b. Instrumentation Amplifier 1NA126P.
c. Metal poll and attachments to hold the force sensor in horizontal position.
d. A piece of thread.
e. Various ¼ W 1% resistors and capacitors.

   2-100 Ω (optional).

   1-332 Ω (optional).

   1-68.1 Ω.

   2-100 nF.
f. 2-2 pin header (optional.)
g. 1–3 pin header (optional).
h. 1–100 Ω trimpot (optional).
i. 1-Force Transducer Fort 100, World Precision Instruments (Figure 9.1).
j. 1–8 pin Din connector CUI SD-80LS.

Software:

- MATLAB®.

S9.6 DETAILED EXPERIMENTAL PROCEDURE

1. The sensitivity of your force transducer, which contains a strain-gauge bridge, is given as 7 μV/V/g by the manufacturer (Fort 100, World Precision Instruments, see Figure 9.1). However, a particular device may have slightly different sensitivity. Thus, we have to measure the transducer's sensitivity first in this studio. The INA126 instrumentation amplifier's gain can be set to any value between five and 10,000 by selection of the resistor value between the pins 1 and 8. R1 sets the gain to 1000. See the table given by the manufacturer below for other gains.

DESIRED GAIN (V/V)	R_G (Ω)	NEAREST 1% R_G VALUE
5	NC	NC
10	16 k	15.8 k
20	5333	5360
50	1779	1770
100	842	845
200	410	412
500	162	162
1000	80.4	80.6
2000	40.1	40.2
5000	16.0	15.8
10,000	8.0	7.87

NC: No Connection.

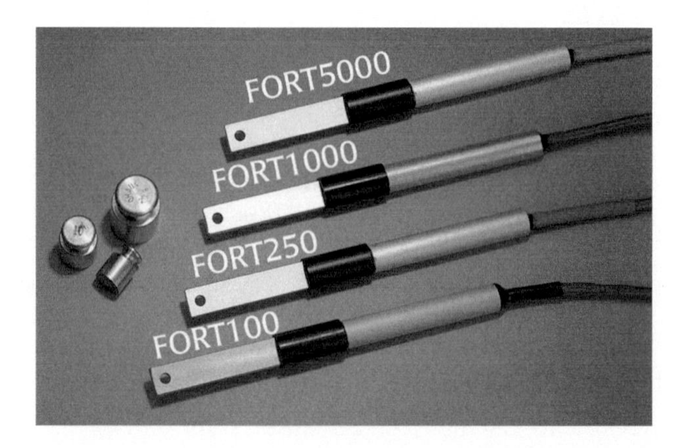

FIGURE 9.1 FORT series force sensors from World Precision Instruments.

2. Build the circuit in Figure 9.2. You will need to solder wires on the connector that can be plugged into the breadboard. Solid 26 AWG wire is recommended.

3. Measure the circuit output with a DC voltmeter and adjust the offset trimpot to make sure that the output is zero for zero force.

4. Next, firmly attach the force sensor to a reference point, such as a pole, about 5–6 inches above the desk. Note the angular position of the transducer rod is important for accurate force measurements. The hole at the tip should be facing the floor.

5. Calibrate the transducer with three weights between 10, 20 and 50 grams. Take a cotton thread and tie one end to the tip of the force sensor through the hole and the other end to the weights, one at a time. Measure the output voltage with a voltmeter, which has better resolution than a typical oscilloscope. Make a plot of output voltage vs. force and plot a linear regression line through the data points (intercept = 0) in MATLAB or Excel. An example is shown in Figure 9.3. Find the slope as the gain of your system in units of V/N (i.e., the calibration value). Is the transducer sensitivity close to what is reported by the manufacturer?

6. Next, tie the other end of the thread attached to the force transducer to the index finger of a subject in your group. Have the subject's elbow and wrist rest on the table while the index finger is pointing horizontally. This will eliminate other sources of tremor from the arm and allow only the finger tremor to be detected by the transducer. Ask the subject to maintain a constant but a minimum level of force.

7. Run MATLAB on the computer and execute "*softscope*" command (or "*analogInputRecorder*" in newer versions of MATLAB). Choose 1000 for sampling frequency and only one channel as input. Set the input voltage range to a minimum but a larger value than the maximum force to be used in this Studio (~1N). Typical values may be ±1 V, ±2.5 V or ±5 V depending on the DAQ board's specifications. Note if the input range selected is too large, the quantization noise in the signal may obscure the small variations in the force signal, as exemplified in Figure 9.4. The quantization noise will also obscure the tremor peaks in the frequency spectrum in the following steps.

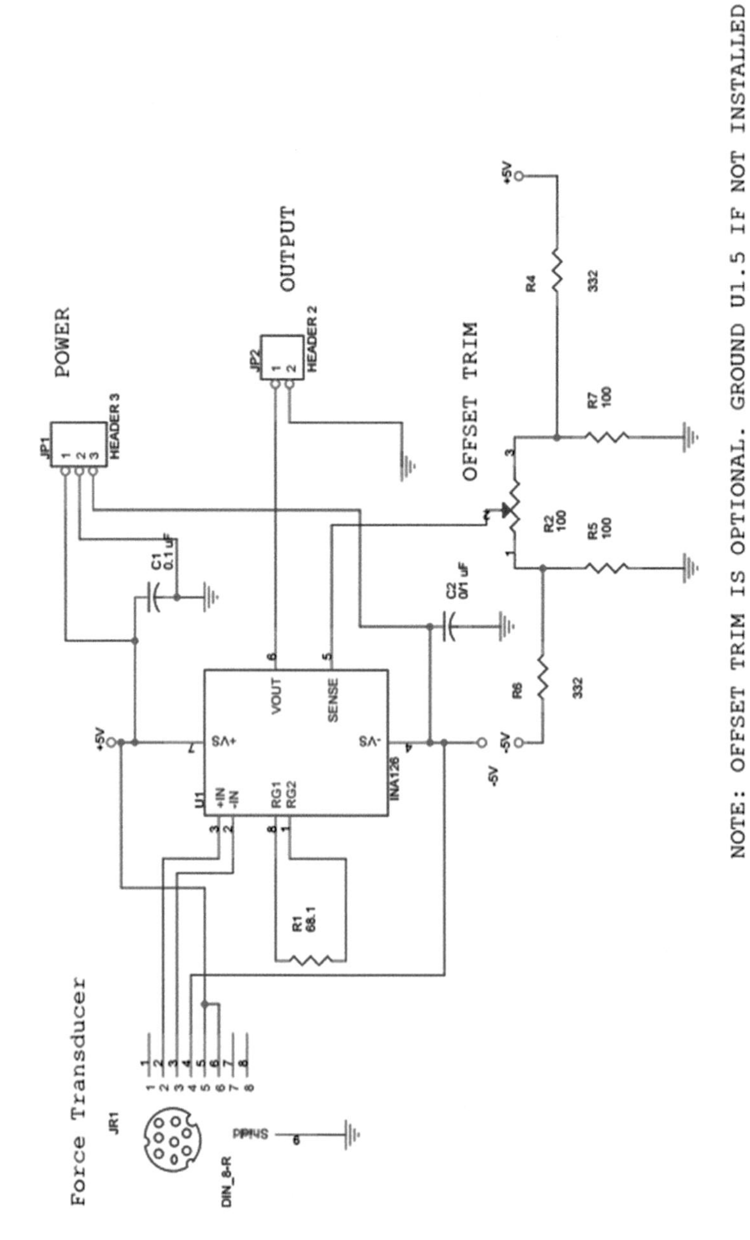

FIGURE 9.2 Force Transducer amplifier schematic. The gain of the instrumentation amplifier is set to ×1000 by R1.

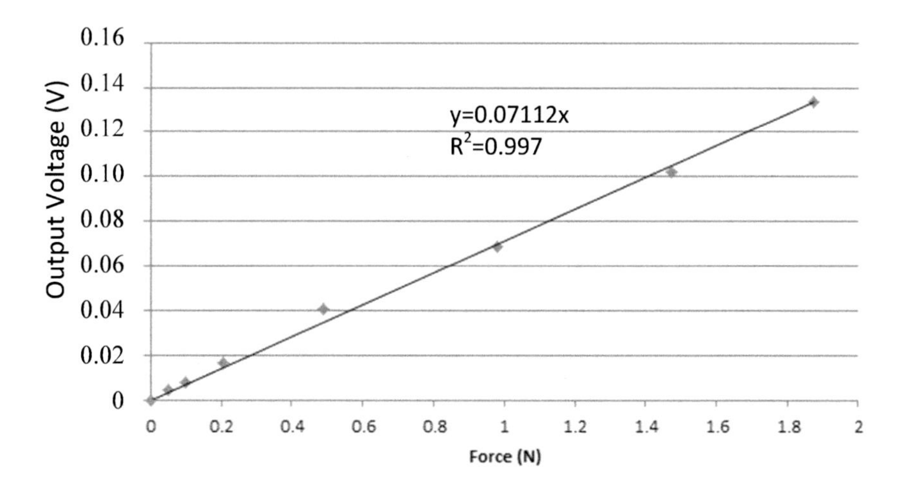

FIGURE 9.3 Force sensor calibration plot. The overall calibration value of the system is the slope of the line, which is 0.7122 V/N. The sensitivity of the force transducer is calculated by accounting for the amplifier gain of ×1000 and the bridge excitation voltage of 10 V. Calibrated sensitivity is 7.112 μV/V/N = (0.07112 V)/(10 V × 1000 × 1 N).

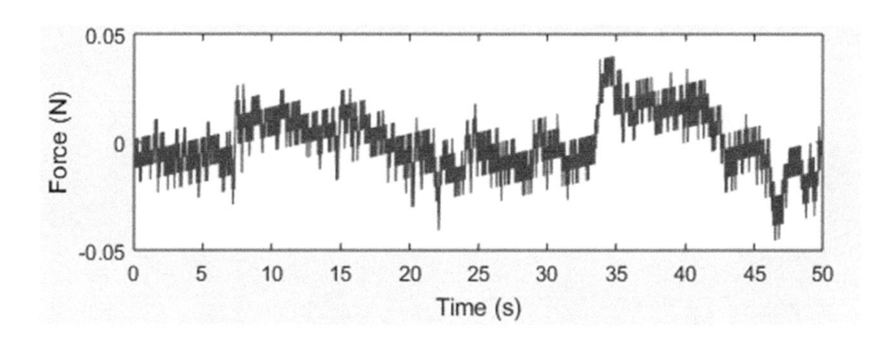

FIGURE 9.4 Force signal with significant quantization noise due to improper selection of the input range for the data acquisition board.

8. Connect the output of your circuit to the input channel of the DAQ board Interface.

9. Collect about one minute's worth of force data under stable conditions without shaking the table (see left panel in Figures 9.5 and 9.6). Make sure that the subject has enough time to attain a constant level of force to avoid any initial jumps in data. You can also cut those large force changes at the beginning or end of the data after saving the data in MATLAB using array commands.

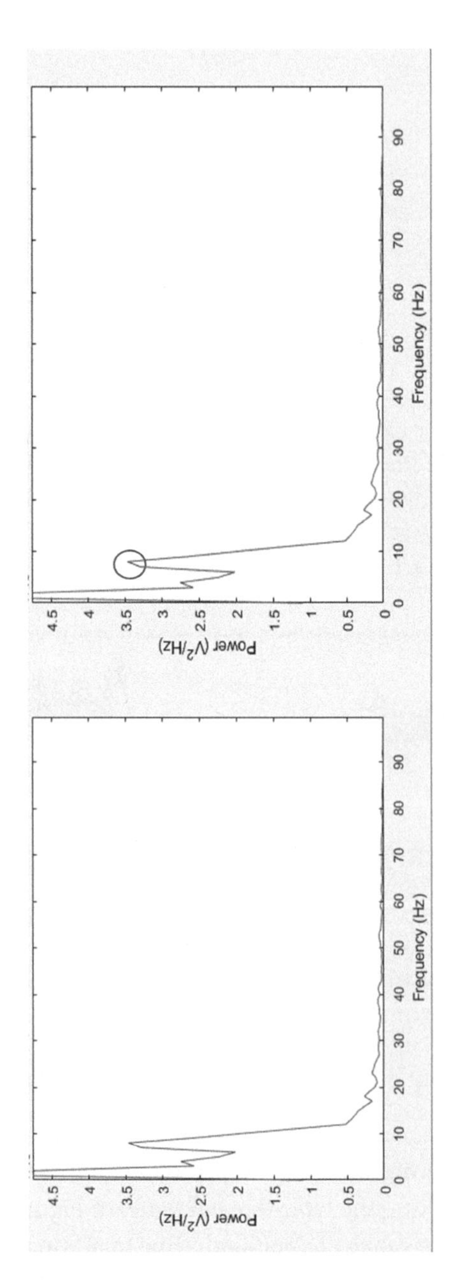

FIGURE 9.5 Force signal collected while the subject made effort to maintain the finger force at a low level (left panel) and its power spectrum (right panel). There is a prominent peak at ~8 Hz.

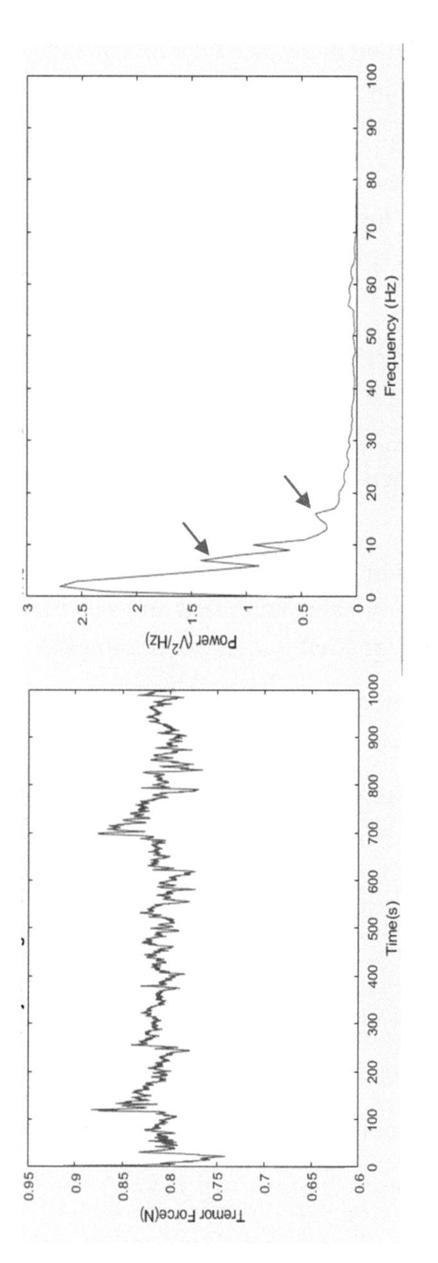

FIGURE 9.6 Force signal collected while the subject tried to keep the finger force at a higher level (left panel) and its power spectrum (right panel). Two additional peaks appear around ~5 Hz and ~15 Hz.

10. Do the power analysis on the data using pmtm(signal, tm) function in MATLAB to investigate the dominant frequencies of tremor (see right panels in Figures 9.5 and 9.6). A typical value for tm is 2–5, which defines the frequency resolution in the plot. Experiment with different values of tm until you are happy with the detail level in the power spectrum.

11. Have the subject get the hand muscles fatigued by making a tight fist and holding for a few minutes. Then, co-contract the finger muscles while applying the force to the rubber band. (i.e., make your finger very tight although applying a small force). Collect another one minute's worth of data while holding the force constant.

12. Do the power analysis on the data using pmtm() function in MATLAB to investigate the dominant frequencies of tremor from the fatigued finger.

13. Repeat the experiments (fatigued and non-fatigued) with other subjects in the group.

14. Plot your signals in units of Newton (N) by dividing the output voltage by the calibration value measured and as a function of time on the horizontal axis, and not in number of samples.

15. Do the dominant frequencies and their power change when finger muscles are fatigued?

16. Do the dominant frequencies and their power change from subject to subject?

In your report, remember to include

In Results:

1. The calibration plot (voltage vs. force).

2. The force plots in time (force vs. time).

3. The power spectra of the force signals (V^2/Hz vs. frequency).

4. The MATLAB codes.

S9.7 CIRCUIT TESTING AND TROUBLESHOOTING

1. Follow the general instruction in previous Studios for troubleshooting the circuit.

2. Confirm that the voltage at the wiper of R2 (middle terminal) can be set to 0 volts.

3. If the output offset cannot be nullified by offset trimpot, the most likely cause is the transducer. The transducer bridge output should be less than a few mV for zero force. A higher voltage output may indicate that the transducer leads are not connected in the proper order to the amplifier. If this is not the cause, the force transducer may be damaged. This can occur if someone has applied excessive levels of force to it.

S9.8 DATA ANALYSIS AND REPORTING

In Theory Section:

1. Information on the force transducer used, FORT 100 (https://www.wpiinc.com/var-2858-large-fort-force-transducer)
2. Physiological information about the causes of tremor.
3. Different types of tremor and neurological disorders that can cause them.

In Results:

1. Include the calibration plot.
2. Include time and frequency plots of the recorded signals.
3. Mark and talk about characteristic frequencies.

In Discussion:

1. Briefly discuss your results of power spectra in comparison to one or two papers you can find in pubmed.org on tremor.
2. Discuss the type of tremor that is observed in this studio.
3. What is Henneman's size principle and how it can be related to tremor?
4. What other sensors/ transducers we could use to detect tremor in the finger?
5. Can a tremor be detected from EMG of the muscles involved?

REFERENCES AND MATERIALS FOR FURTHER READING

References marked with an asterisk (*) are recommended to those interested in expanding on the content of this chapter.

1. *INA126 Datasheet*, Texas Instruments, SBOS051D, rev. January 2018.
2. *FORT 100 Datasheet FORT100etc_IM2*, World Precision Instruments Inc.
3. *Brown, T.I.; Rack, P.M.; Ross, H.F., Different types of tremor in the human thumb brown, *J. Physiol.* (1982) 332:113–123.
4. *McAuley, J.H.; Rothwell, J.C.; Marsden, C.D., Frequency peaks of tremor, muscle vibration and electromyographic activity at 10 Hz, 20 Hz, and 40 Hz during human finger muscle contraction that may reflect rhythmicities of central neural firing, *Exp. Brain Res.* (1977) 114:525–541.
5. *Adrian, E.D.; Moruzzi, G., Impulse in the pyramidal tract, *J. Physiol.* (1939) 97:153–199.

Studio 10

Optical Isolation of Physiological Amplifiers

S10.1 BACKGROUND

When designing medical equipment, it is essential, and required by regulation, to keep patient safety in mind with regard to the potential for electric shock hazards. With any AC powered (mains connected) electrical device it is possible for a failure to occur that can apply mains voltage to the patient resulting in electric shock. Regulation requires multiple means of patient protection (MOPPS) and means of operator protection (MOOPS) to insure safety. In this studio, we become familiar with optical isolation, which allows two electrical isolated circuits to pass a signal from one to the other using light to transmit the signal while the circuits are completely electrically isolated. The students will also become familiar with DC to DC isolation that provides power from a mains-connected circuit to the isolated circuit.

S10.2 OVERVIEW OF THE EXPERIMENT

This studio will implement a galvanically isolated amplifier that can be used to transmit physiological signals derived from a patient connection to a non-isolated oscilloscope and PC (personal computer). Specifically, in this experiment we will use a three-lead electrocardiogram (ECG) as the physiological signal.

S10.3 LEARNING OBJECTIVES

The objectives of this studio are to:

- Understand optocouplers and their limitations.
- Understand galvanic isolation.
- Understand isolated DC to DC power supplies.

S10.4 SAFETY NOTES

It is extremely important that the subject is not connected directly to the ground of any of the instruments used during this studio – e.g., oscilloscope, voltage supply, computer. In hospitals the clinical equipment has isolated grounds that prevent the passage of AC or DC current from such equipment to the patients. Isolated grounds are a safety measure to protect patients in case of faulty equipment. In the worst situation, a direct connection between the subject and the ground of a faulty apparatus may result in a fatality. All electric equipment in the laboratory should be tested for safety prior to usage. This is normally the responsibility of the host institution.

S10.5 EQUIPMENT, TOOLS, ELECTRONIC COMPONENTS AND SOFTWARE

Additional information about the use of the items required in this studio can be found in the Introduction and in the Appendices.

Equipment:

- Breadboard.
- Power supply (+12 V).
- Multimeter.
- Oscilloscope.
- Data Acquisition Board (DAQ).
- Computer.
- AC Lab Amplifier.

- ECG cable for above.
- ECG electrodes.

Tools:

- BNC-to-microclip cables.
- Jumper wires for the breadboard.

Electronic Components (remember to download the available datasheets):

a. 1 Opto-coupler Vishay IL300-G.

b. 2 Operational Amplifiers, LM358AP.

c. 1 DC-DC power supply XPPower IA1205S.

d. Various ¼ W resistors and capacitors (low leakage preferred).

1–100K.

3–10 K.

1–47 K.

1– 6.8 K.

1– 270.

3–0.1 μF.

1–0.001μF.

e. 3–2 pin header Harwin M20-9990246 (optional).

Software:

- MATLAB®.

S10.6 CIRCUIT OPERATION

The circuit is divided into two parts (Figure 10.1), the left-hand side is the isolated side and the right side is the non-isolated side. Power is provided to the circuit by applying 12 volts to the non-isolated side where it powers the Op-Amp, U2, the DC-DC converter, T1 and the phototransistor in optoisolator A1. On the isolated side, the output of T1 provides ±5 volts

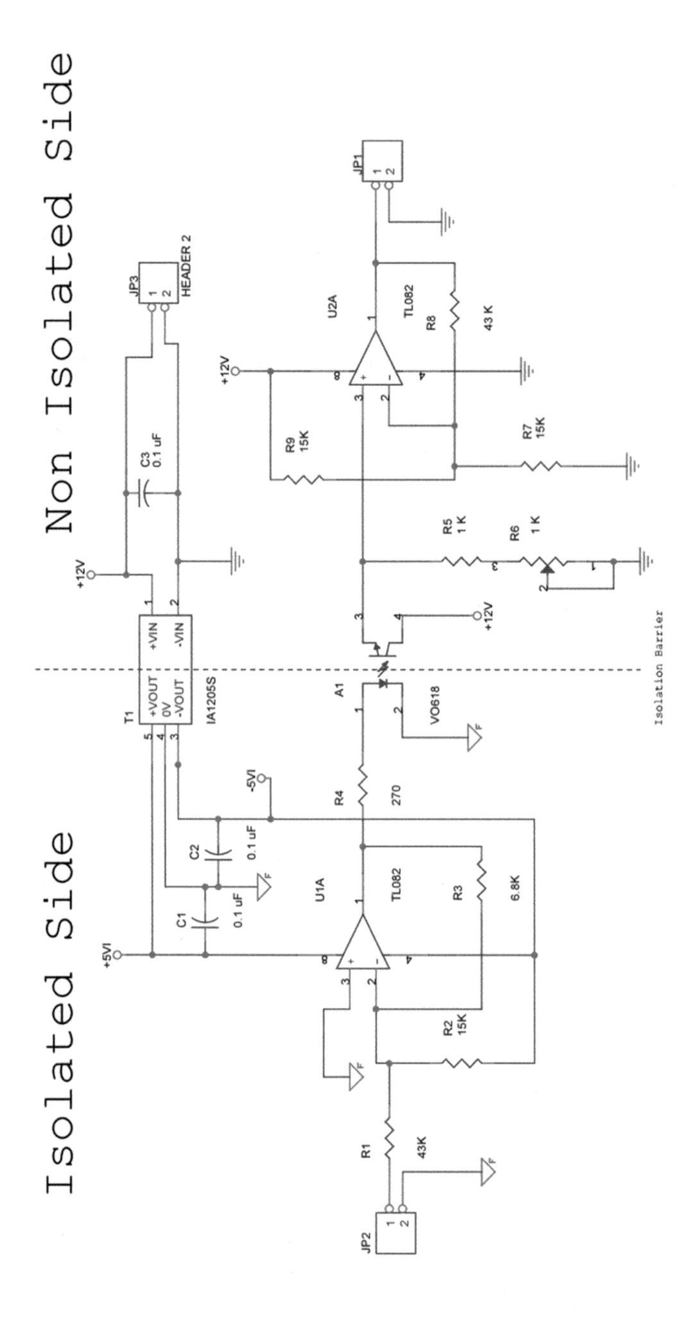

FIGURE 10.1 Schematic diagram for the isolation amplifier.

to Op-Amp U1. It isolates the signal ground from the floating ground by using a small transformer that has an 80 kHz drive on its primary, and rectifying and filtering the secondary outputs to provide DC power. The ±5 volt DC outputs power the Op-Amp U1. U1 drives the photodiode in A1 so that the quiescent current is 4.3 mA by setting the output voltage of U1 to 2.3 volts. The ±1 volt signal applied to JP2 will modulate the 2.3 volt output of U1 with a ±0.16 signal causing a variation in the light output of the photodiode while keeping the variation in the light output in a quasi linear operating region.

On the non-isolated side, the output of the phototransistor applies current to R5/R6 which is adjusted to a nominal DC value of 6 volts. U2's gain is set to reproduce the same nominal signal level as at the opto-isolator circuits input.

S10.7 DETAILED EXPERIMENTAL PROCEDURE

1. Build the circuit given in the schematic on a breadboard.
2. Compare the circuit with the schematic and verify all the connections before applying the power supply.
3. With your ohmmeter confirm that there is no connection between the floating ground and the signal ground.
4. Apply 12 volts DC to the breadboard. Adjust R6 to that U2 pin1 is 6 volts ± 0.5 volts.
5. Apply a 10 Hz sinusoidal signal with ±1 V (peak-to-peak) amplitude from the signal generator to the input of the circuit (with respect to ground) to simulate an electrophysiological signal.
6. Now observe the output of the second Op-Amp on the oscilloscope (Channel 1) as well as the original sinusoidal signal from the signal generator (Channel 2) and compare if the waveforms are identical, if there are any distortions in the output waveform, and if there is any phase shift.
7. Test your circuit at frequencies from 0.05 Hz up to 1 MHz increasing by decades (e.g. 0.05, 1, 10, 100, 1000, 10,000, 100,000, 200,000, 300,000, ….. 1,000,000 Hz) while making amplitude and phase shift measurements on the oscilloscope.

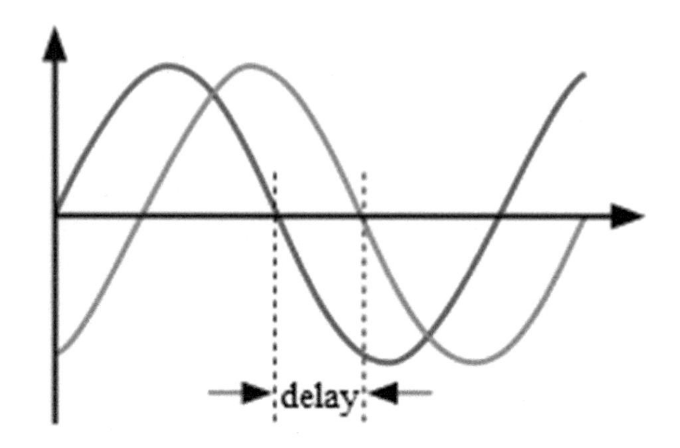

$$\text{Delay}(\text{ms}) = \Delta t = \tau$$

$$T\frac{1}{f}$$

$$\text{Phase} = -\frac{\tau}{T} \times 360°$$

8. Make a plot of the output amplitude and phase as a function of frequency in Excel or MATLAB. Use log-log scale for the axes.
9. Do a Fast Fourier transform (FFT) on the oscilloscope using the Math function (if available) on the output signal. Is there distortion? How do you tell if there is distortion from the FFT? (Hint: harmonics.)
10. Collect a few seconds worth of data into MATLAB using a function such as *analogInputRecorder*. And do an FFT on the output data in MATLAB.
11. Quantify the total harmonic distortion (THD) by taking the sum of squares of the higher harmonics of the original signal and dividing it by the square of the first harmonic (fundamental frequency).

$$\text{THD} = \frac{\sum_{n=1}^{\infty} \text{Mag}_n^2}{\text{Mag}_{\text{fundamental}}^2}$$

12. What is the amount of THD in percentage?

13. Compare your results with the sample results by other students shown at the end in Figures 10.2–10.4. (Note that the corner frequency that you are building will probably be different.)

Bonus work if there is time:

14. Now, let us use this optical isolation for ECG signals. Get an ECG cable for the AC amplifier and connect it to one of the subjects in the group.
15. Use sufficient amplification and appropriate filtering for ECG (0.01–100 Hz) to see a clear ECG waveform on the oscilloscope.
16. Connect the output of the amplifier to the input of your circuit and then display the output on the oscilloscope.
17. Does the signal look clean and undistorted? Is it saturated? How would you get rid of saturation?

In your report, remember to include:

1. Introduction: Theory on optocouplers, photodiodes and DC-DC isolated power supplies.

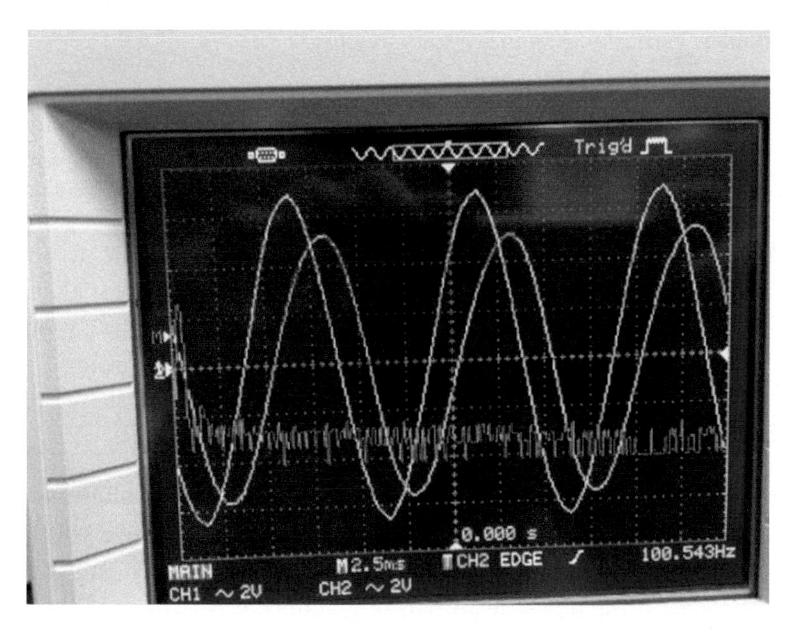

FIGURE 10.2 Sample signals. Input (yellow) and Output (blue) signals are shown with a phase shift. FFT of the input signal is in red with a peak at 100 Hz.

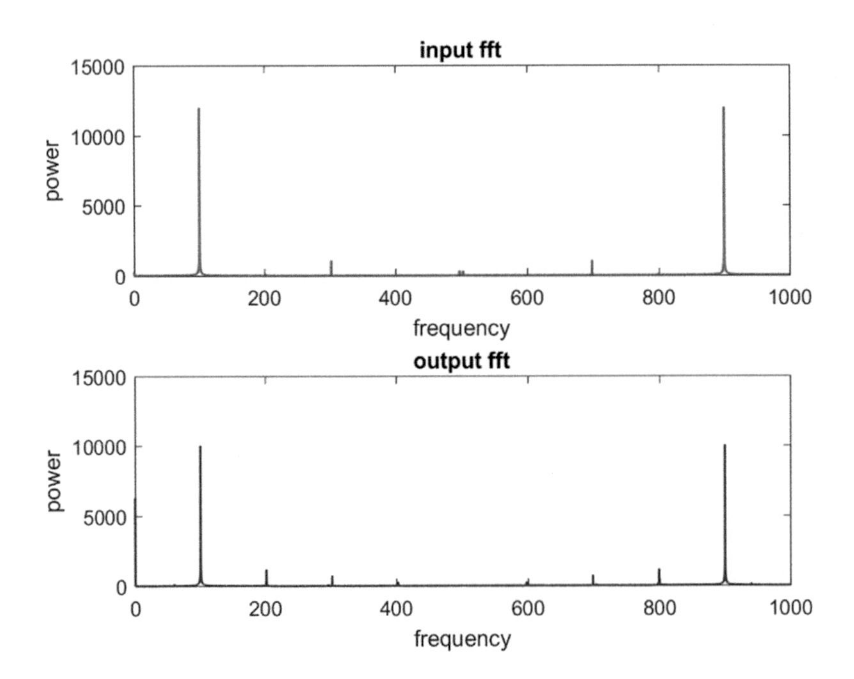

FIGURE 10.3 Double-sided FFT of the input and output signals collected at a sampling rate of 1000 samples per second. The fundamental frequency is 100 Hz. Both the input and output signals have harmonics, but with different relative amplitudes. Vertical scale is in arbitrary units.

2. Results: Plot the waveforms at several frequencies. Plot the amplitude and phase of the transfer function using a log-log scale for the axes and dB units for the amplitude.

3. Discussion: Discuss the problems encountered and how they were solved in building and testing of the circuit. Discuss potential usages of the circuit for electrophysiological signals. Discuss limitations of the circuit in terms of frequency, noise and signal amplitude.

S10.8 CIRCUIT TESTING AND TROUBLESHOOTING

1. Before turning the power on, carefully check if the voltage supply is applied to the circuit with the correct polarity. Using a voltmeter with one lead connected to the ground, check the 5 V and ground voltages at the power terminals of all chips. It is a time saving practice to do this as the first step of troubleshooting in any circuit, analog or digital.

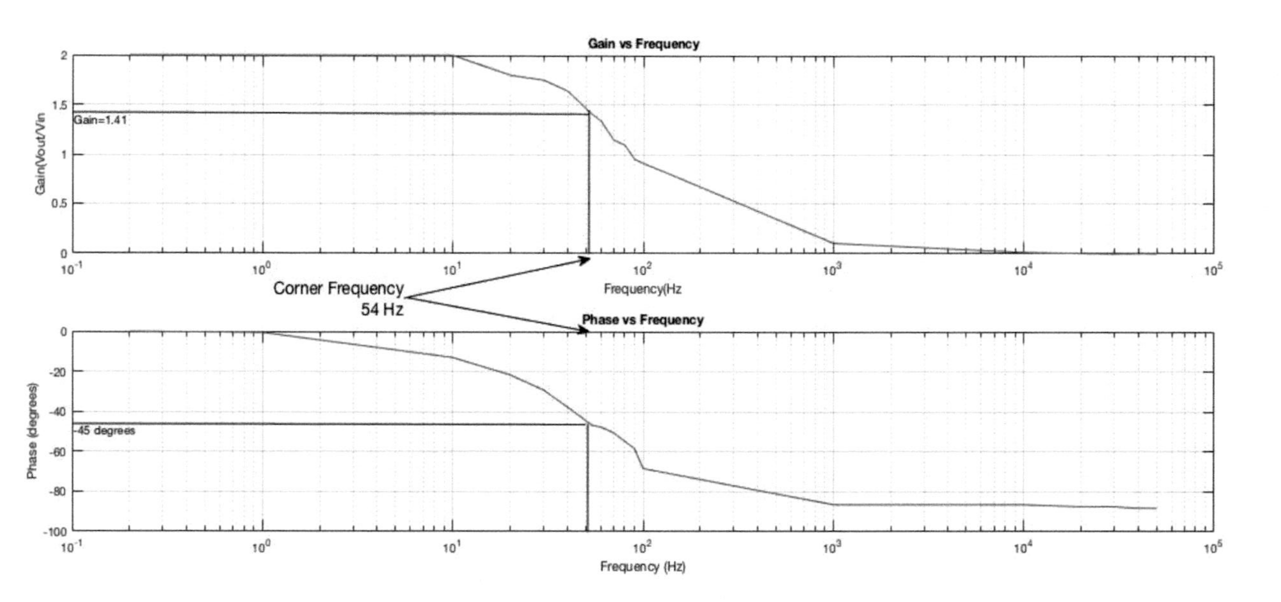

FIGURE 10.4 Magnitude and phase plot of an optical isolation unit built by students using 4N33 optocoupler. The corner frequency is 54 Hz, marked at –3dB (or 0.707 of the maximum gain) on the magnitude plot and –45 degree on the phase.

2. Another good practice is to measure the total current that the circuit is drawing from the power supply (most modern power supplies have a voltage and a current display). If the current is higher than a few tens of a mA, or one of the chips is becoming too warm (be careful not to burn your fingers) you may conclude there is a broken chip or a short circuit due to an incorrect connection.
3. Confirm that ±5 Volts is available at the output of T1.
4. With no signal at the input, check the current in R4, it should be 4 ± 1 mA.

S10.9 QUESTIONS FOR BRAINSTORMING

a. What happens if you increase the gain of U1?
b. What happens if you increase/decrease the quiescent LED current?
c. What happens if you misadjust R6?
d. What happens if you connect the floating ground to the signal ground?

S10.10 IMPORTANT TOPICS TO INCLUDE IN LAB REPORT

a. The need for galvanic isolation.
b. The need for creepage and air gap distances between the two grounds.
c. Why analog optical isolation (as opposed to digital optical isolation) may not be the best method of signal transmission across the barrier.

REFERENCES AND MATERIALS FOR FURTHER READING

References marked with an asterisk (*) are recommended to those interested in expanding on the content of this chapter.

1. *LM358 Datasheet*, Texas Instruments, SLOS068U, January 2017.
2. *IL300 Vishay Datasheet*, Document Number: 83622 Rev. 1, 8 June 2014.
3. *IA Series Datasheet*, XPPower, January 21, 2011.
4. **Application Note AN 2012-10 V1.0*, Infineon Technologies Austria AG, October 2012.

STUDIO 11

Extraction of Respiratory Rate from ECG (ECG-Derived Respiration-EDR)

S11.1 BACKGROUND

It is desirable to minimize the number of electrodes, sensors and cables attached to a patient. EDR allows for respiration to be derived from the electrocardiogram (ECG) signal, eliminating the need for a separate respiration sensor and cable. EDR is possible because the ECG electrodes on the chest move relative to the heart, varying the transthoracic impedance as the lungs expand and contract with each breath. Although this method does not provide a calibrated respiratory waveform, it is useful for determining the respiratory rate.

S11.2 OVERVIEW OF THE EXPERIMENT

In this experiment a physiological amplifier is used to amplify and condition the ECG signal. The output of the amplifier will be analyzed in the MATLAB® environment for calculation of the respiration rate.

S11.3 LEARNING OBJECTIVES

The objectives of this studio are to:

- To learn how to acquire physiological signals and process them in MATLAB®.

S11.4 SAFETY NOTES

It is extremely important that the subject is not connected directly to the ground of any of the instruments used during this studio – e.g., Grass Amplifier or computer. In hospitals the clinical equipment has isolated grounds that prevent the passage of AC or DC current from such equipment to the patients. Isolated grounds are a safety measure to protect patients in case of faulty equipment. In the worst situation, a direct connection between the subject and the ground of a faulty apparatus may result in a fatality.

S11.5 EQUIPMENT, TOOLS, ELECTRONIC COMPONENTS AND SOFTWARE

Additional information about the use of the items required in this studio can be found in the Introduction and the Appendices.

Equipment:

- AC physiological amplifier (or any ECG Amplifier for human subjects).
- Computer with a Data Acquisition Board (DAQ).

Cables and sensors:

a. Disposable ECG electrodes.

b. Respiband from Grass (if using Grass Amplifier).

c. ECG electrode gel (in case electrodes become dry).

Software:

- MATLAB®.

S11.6 DETAILED EXPERIMENTAL PROCEDURE

Data Collection:

1. Use the AC Grass Amplifiers (rated for human subjects) for both the ECG and respiratory measurements and MATLAB to collect the data (e.g., *analogInputRecoder* function).

2. For ECG, set the Grass amplifier gain initially to 10,000 and adjust it later to get a clear view of the signal on the screen. High gains produce better signal-to-noise ratios (SNR), as long as the amplifier output does not saturate. Set the amplifier filters to low cut-off = 0.3 Hz and high cut-off = 100 Hz. 60 Hz notch filter should be IN.
3. Identify a subject in your group and attach two disposable electrodes on their wrists and connect these two electrodes to the differential inputs of the amplifier using a Grass amplifier input cable. Connect a third electrode as a reference to the left leg (or on the abdomen if more convenient) and then to the common terminal of the amplifier cable.
4. For the respiratory measurements, wrap the Respiband around the chest of the subject and connect the two output leads to the differential inputs of the second AC Grass Amplifier. Set the gain to 100, the low cut-off = 0.03 Hz and high cut-off = 30 Hz. 60 Hz notch filter should be IN.
5. Connect the amplifier outputs to the inputs of the data acquisition card using BNC cables and run *analogInputRecorder* in MATLAB. Use 1000 samples/sec for the sampling rate and the smallest possible input range in order to maximize the resolution (but without amplitude saturation).
6. Collect 60,000 data points from each channel simultaneously while the subject is breathing normally for one minute's worth of data. Export the signal into MATLAB. The subject should not move during data collection to ensure stable signals.
7. Repeat the measurement two more times at faster and slower respiratory rates and export the data to MATLAB under different variable names.

S11.7 DATA ANALYSIS AND REPORTING

1. First, filter your signal with appropriate high-pass filters to get rid of any baseline wandering. Use a fourth order Butterworth (*butter() in MATLAB*) filter and filter the signal using *filtfilt()* function to increase the sharpness of the filter while avoiding any time delays that the filtering might cause. You need to experiment with different corner frequencies to find one that removes the baseline wandering but not the slow components of the ECG waveform such as the P and T waves. Optimum corner frequency will be around 0.5 Hz. (You may use the attached m.file for filtering signals.)

2. If the signal is noisy, filter it also with a low pass to remove the noise. Experiment with different corner frequencies between 30–100 Hz to find a proper corner frequency that removes most of the noise but does not attenuate the QRS wave.

3. Now, find the peak points of the QRS waves using the *findpeaks()* command in MATLAB. You may set a threshold to detect the peaks or use other options available with this command in different versions of MATLAB.

4. Plot the ECG signal again and place a red asterisk at the QRS peaks to verify that the peaks were detected correctly.

5. Next, find the zero-crossing points on each side of the QRS peak for each heartbeat and save their time indices in a variable.

6. Integrate the areas underneath the QRS peaks (just sum the numbers between the zero crossing points for each QRS) and plot the QRS area as a function of time.

7. Resample the collected respiratory signal at the times of the QRS peaks detected in a previous step and put these values into a new array.

8. Now the new respiratory array and the integrated QRS should have the same number of points. Plot them both on the same figure in different colors and calculate their correlation using *corrcoef()* command. Is the correlation high? What is the p-value given by the correlation as an output variable? A p-value smaller than 0.05 indicates that the correlation value computed has a strong statistical power. Does the ECG derived respiratory signal (areas under the QRS waves) look similar to the actual respiratory signal as a function of time?

9. Apply this analysis to other recordings you made under fast and slow breathing conditions and make records of the correlation and p-values.

10. As a second approach, instead of using the area of the QRS wave, only use the QRS peak amplitudes to derive the respiratory signal (as exemplified in the attached m.file). Apply this method to all collected signals and determine which method gives you higher correlation with the actual respiratory signals.

11. Compare your results with those sample data shown in Figures 11.1–11.3.

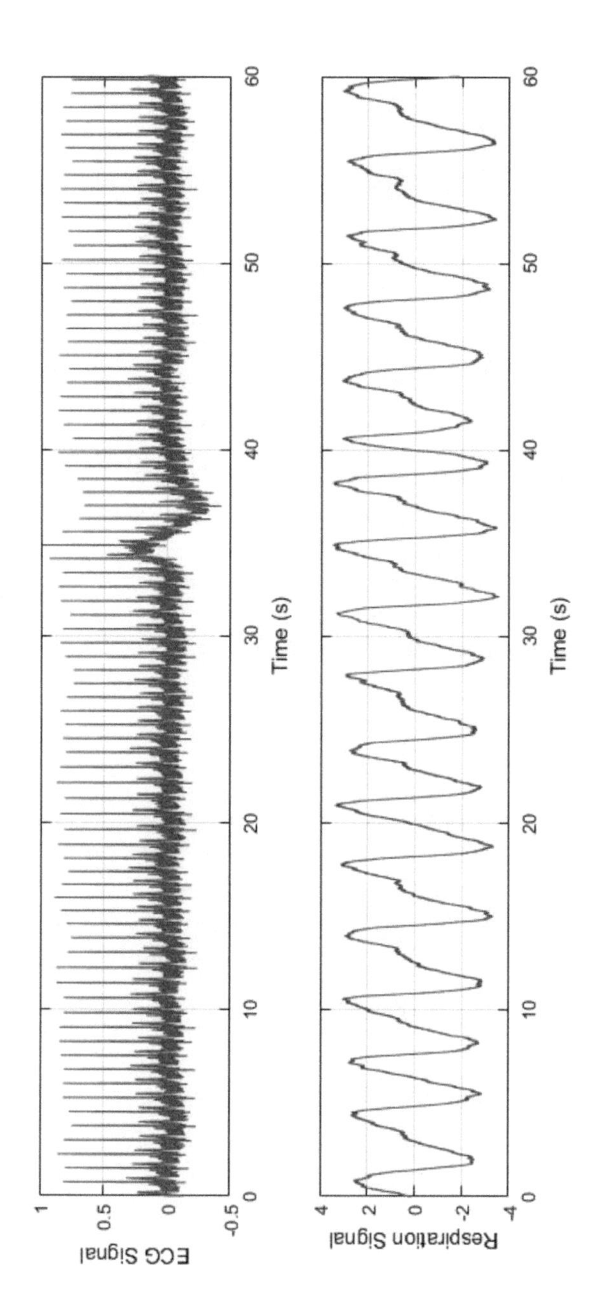

FIGURE 11.1 Raw ECG and respiratory signals collected for a duration of 60 s by students. Notice the movement artifact around t = 35 s. This can be removed with a digital high-pass filter (e.g., Butterworth filter, *butter( )* in MATLAB) with a corner frequency of about 0.5 Hz on the computer.

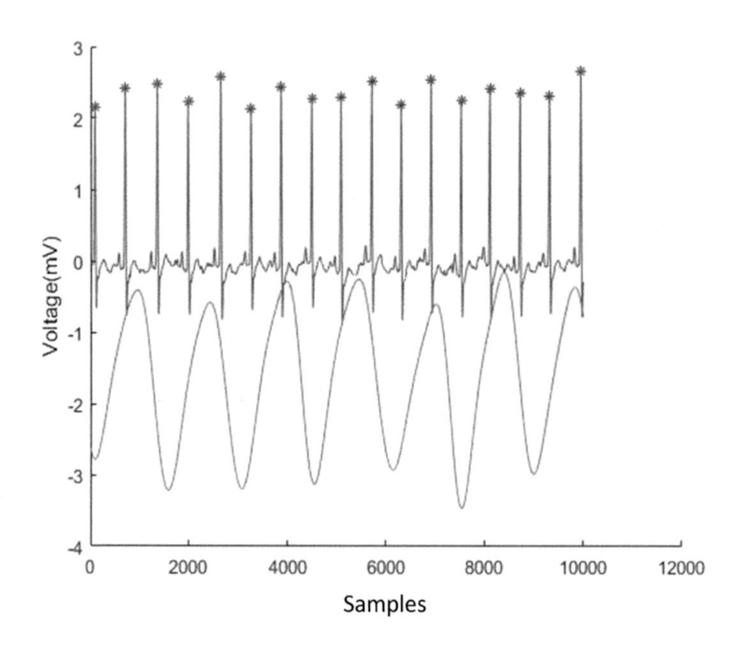

FIGURE 11.2 Filtered ECG and respiration signals during slow breathing. Sampling rate is 1000 samples per second. ECG peaks are found and marked with asterisks. The correlation coefficient between the ECG peaks and the respiration signal is 0.86 (p <0.005). Note the phase delay between the two signals.

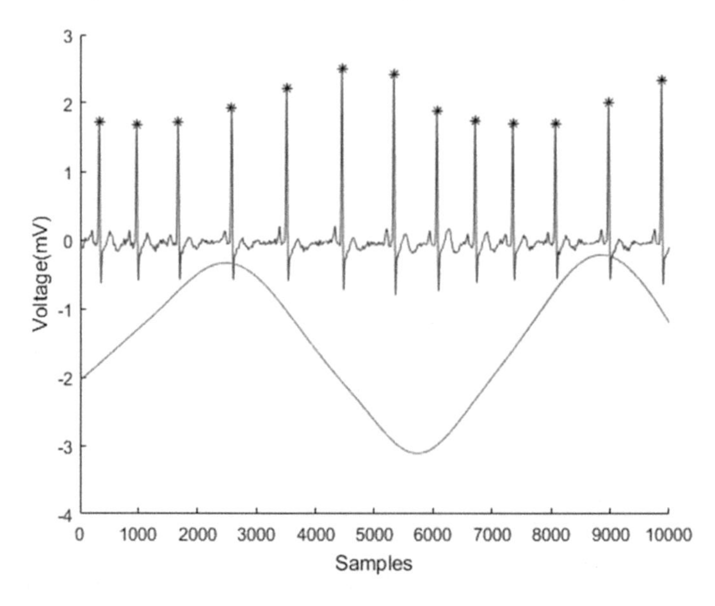

FIGURE 11.3 Filtered ECG and respiration signals during fast breathing. The correlation coefficient between the ECG peaks and the respiration signal is 0.72 (p <0.005).

12. Notice the large phase delay in Figure 11.2. Find the average delay in your signals and compute the correlation coefficient again after correcting for this delay. (Hint: use *xcorr* function in MATLAB). Is the correlation coefficient higher now? Explain why.

REFERENCES AND MATERIALS FOR FURTHER READING

References marked with an asterisk (*) are recommended to those interested in expanding on the content of this chapter.

1. Moody, George B.; Roger G. Mark; Andrea Zoccola; and Sara Mantero. Derivation of respiratory signals from multi-lead ECGs, *Computers in Cardiology* (1986) 12:113–116, IEEE Computer Society Press.
2. Moody, George B.; Roger G. Mark; Marjorie A. Bump; Joseph S. Weinstein; Aaron D. Berman; Joseph E. Mietus; and Ary L. Goldberger. Clinical validation of the ECG-derived respiration (EDR) technique, *Computers in Cardiology* (1986) 13:507–510, IEEE Computer Society Press.

Studio 12

Heart Rate Variability Analysis in Frequency Domain

S12.1 BACKGROUND

The twentieth century, with the generalized use of electrocardiography (ECG or EKG) in clinical settings, marked the start of the quest to unravel the relationship between heart rate variability (HRV) and disease [1]. Multiple cardiovascular risk factors and disease states have been found to reduce HRV, including diabetes, smoking, obesity, work stress, hypertension and heart failure [2]. For example, patients recovering from a myocardial infarction show reduced HRV, with the largest reductions in HRV associated to the highest risk for sudden death of those patients [3]. The details of the physiological basis of HRV remain an open question and are the subject of intensive research [2]. Currently the consensus is that multiple factors determine HRV, including the parasympathetic and sympathetic divisions of the autonomic nervous system as well as the mechanical stimuli associated to the recurrent filling of the lungs (with air) and the heart (with blood).

A number of parameters have been defined to analyze HRV and they can be organized into two main categories: time-domain and frequency-domain parameters [3]. Pertaining to the time domain, the *NN time interval* quantifies the time between two adjacent *normal* heartbeats, which are defined as QRS complexes (Studio 2) produced by the sinoatrial node (SA node) as opposed to those produced by atrial or ventricular arrhythmias. The *SDNN* parameter computes the variability in a sequence of NN time intervals by calculating the standard deviation of the NN time intervals. The *average heart rate* results of dividing a number of heartbeats (n) by the time interval during which those n beats occurred. If the number of recorded heartbeats is larger than the chosen n, one can use the latest n cardiac cycles to determine a new average after each heartbeat; this type of continuously calculated average is called the *running average heart rate*. The significance of average heart rates in the study of HRV, however, is limited because the averaging process – especially with large values of n – may mask temporary or intermittent changes in the beat-to-beat intervals (also known as inter-beat or R-R intervals) that are key components of HRV. Alternatively, the *instantaneous heart rate* (or beat-to-beat heart rate) can be calculated by taking the reciprocal of the time interval between two consecutive heartbeats. The variability of the heart rate can also be expressed in terms of frequency such that rapid (or slow) changes in heart rate are described by high (or low) frequencies. Four frequency bands are typically identified in the spectrum of the HRV of an individual at rest: *ultra-low frequency* (ULF, <0.003 Hz), *very low frequency* (VLF, 0.003–0.04 Hz), *low frequency* (LF, 0.04–0.15 Hz) and *high frequency* (HF, 0.15–0.4 Hz). Researchers have linked bands of the HRV spectrum to the activity of divisions of the autonomic nervous system although such understanding still lacks the consensus of the HRV community [1].

The study of HRV requires equipment that is capable of recording the heartbeat continuously for some periods of time (for example, detection of VLF in HRV requires recordings of at least 24 h) and without causing distress to the subject under study. The simplest device available for studying HRV is the *photoplethysmograph* (PPG), which uses light to detect the changes in volume caused by the cardiac cycle in tissues highly permeated by blood capillaries such as fingertips, toes, or ear lobes (see Studio 8). It is worth noting that ECGs typically provide more detailed information of the heartbeat than the PPG, which may be necessary for complex diagnosis of patients with arrhythmias.

Heart rate variability (HRV) is the cycle-to-cycle variation in the interval between heartbeats. There are several ways of measuring this variability, in this studio we will use the frequency domain approach on the ECG signal to measure the HRV. Using spectral analysis, the HRV can be broken down into the following components:

ULF is the power density number for the ultra-low frequency range (<0.003 Hz), and its prognosis of sudden cardiac death taken from 24-hour ECG recordings is highly accurate.

VLF is power density number for the very low-frequency range (0.003–0.04 Hz), and it is thought to be connected to thermoregulation, the renin-angiotensin system, and changes in physical activity.

LF is the power density number for the low-frequency range (0.04–0.15 Hz) that is generated mainly by sympathetic activity. It is hypothesized that baroreceptor modulation is a major component of LF power.

HF is the power of the high-frequency zone (0.15–0.40 Hz) and is derived from vagal activity which is modulated by respiration.

Since LF represents mainly sympathetic activity and HF represents vagal activity, the ratio (LF/HF) is a good indicator of sympathetic-vagal balance. This ratio is used to assess the balance of the autonomic nervous system in various diseases.

S12.2 OVERVIEW OF THE EXPERIMENT

This studio we will first collect ECG waveforms for spectral analysis (Fast Fourier Transform – FFT). Note that the plethysmograph data collected in Studio 8 can be used here for HRV analysis to substitute the ECG signals. If plethysmograph data to be used, one can skip the ECG collection part and move to data analysis below.

S12.3 LEARNING OBJECTIVE

The objective of this studio is to learn how to do Heart Rate Variability Analysis in the frequency domain using ECG Signals.

S12.4 SAFETY NOTES

It is extremely important that the subject is not connected directly to the ground of any of the instruments used during this studio – e.g., oscilloscope, voltage supply, computer. In hospitals the clinical equipment has isolated grounds that prevent the passage of AC or DC current from such equipment to the patients. Isolated grounds are a safety measure to protect patients in case of faulty equipment. In the worst situation, a direct connection between the subject and the ground of a faulty apparatus may result in a fatality.

S12.5 EQUIPMENT AND SOFTWARE

Additional information about the use of the items required in this studio can be found in the Introduction and the Appendices.

Equipment:

a. Data Acquisition Board (DAQ).

b. Computer.

c. ECG Amplifier.

d. ECG cable for the above amplifier.

e. ECG electrodes.

Software:

- MATLAB®.

S12.6 DETAILED EXPERIMENTAL PROCEDURE

1. Set the ECG amplifier gain initially to 10,000 and adjust later to get a clear view of the signal on the screen. High gains usually produce better signal-to-noise ratios (although this may not be intuitive), as long as the amplifier output does not saturate. Set the amplifier's filters to low cut-off = 0.3 Hz and high cut-off = 100 Hz. The 60 Hz notch filter should be turned on. Use 1000 samples/sec for the sampling rate on the computer. (Note that the sampling rate can be reduced as low as 200 samples/s if desired. A high sampling rate provides precise detection of QRS peak times.)

2. Identify a subject in your group and attach two disposable electrodes on the arms or shoulders and connect these two electrodes to the differential inputs of the amplifier.

3. Connect a third electrode as a reference to the left leg (or on the abdomen if more convenient) and then to the common terminal (COM) of the amplifier.

4. Using this method, collect about a 100 s worth of ECG signal while relaxed (sitting quietly, maybe eyes closed).

5. Save 100,000 data points for 100 s worth of data in each trial.

6. Load the signal into MATLAB.

7. Next, ask the subject to exhale against a closed mouth and nose using moderate force, as if blowing up a balloon. This is called Valsalva maneuver and it is known to increase variability in the heart rate. Collect another set of data for a 30 s Valsalva maneuver for comparison.

S12.7 DATA ANALYSIS

1. Take a look at the power spectrum of the ECG signal using the pmtm() or pwelch() functions in MATLAB. Decide where the ECG signal and where the movement artifacts are at the lower end of the spectrum. A good separation point is usually about 0.5 Hz.

2. Use the butter() function in MATLAB to design a high-pass fourth order Butterworth filter with corner frequency of fc = 0.5 Hz to suppress the baseline wandering and other low frequency artifacts while emphasizing the QRS waveform. Detrend () function can be used before filtering to remove gradual shifting in the baseline during recording. Use filtfilt() function to filter the signal with the coefficients of the Butterworth filtered designed. If the filtering does not emphasize the QRS sufficiently, vary the filter corner frequency around 0.5 Hz (based on the power spectrum) to maximize the signal-to-noise ratio.

3. From the power spectrum, decide if a low-pass filter, for instance around 30 Hz, is needed to remove high-frequency noise. Design another Butterworth filter and apply if necessary. The low-pass filter should not attenuate the high frequency components of the ECG waveform, mostly present around the fast-changing QRS component. If QRS peak decreases after filtering, increase the corner frequency of the low pass.

4. Use the findpeaks() function in MATLAB (with "MinPeakDistance" and "MinPeakHeight" options) to find the locations of the QRS peaks in time. Make sure you only detect the QRS peaks and not other local maxima due to noise or artifact. Plot the ECG signal with the QRS peaks marked with a red asterisk (as in Figures 12.1 and 12.2 with and without Valsalva maneuver respectively).

5. Calculate the Inter-Beat-Intervals (IBI) by finding the time differences between the QRS peaks (in units of seconds) and place them in an array.

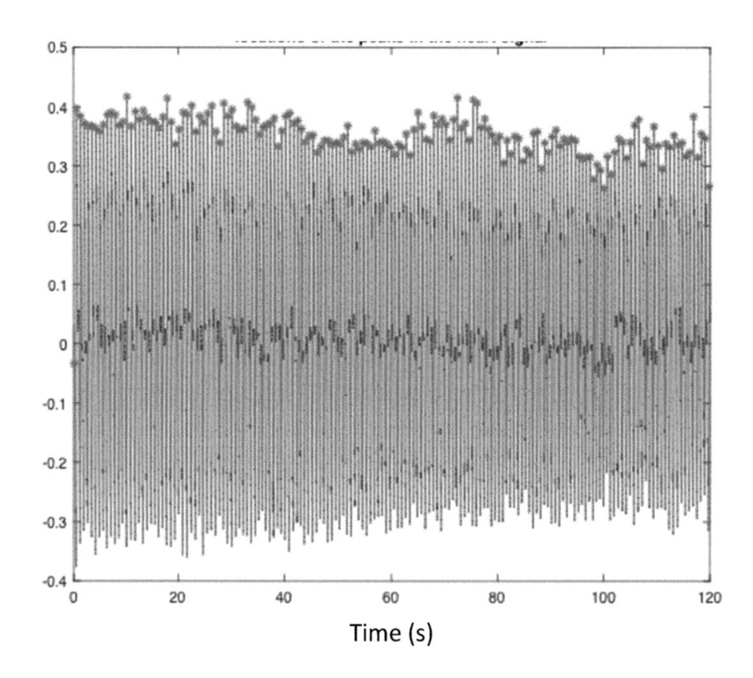

FIGURE 12.1 Band-pass filtered 120 s long ECG data with the QRS peaks found using findpeaks() function in MATLAB and marked with asterisks. The subject is breathing normally.

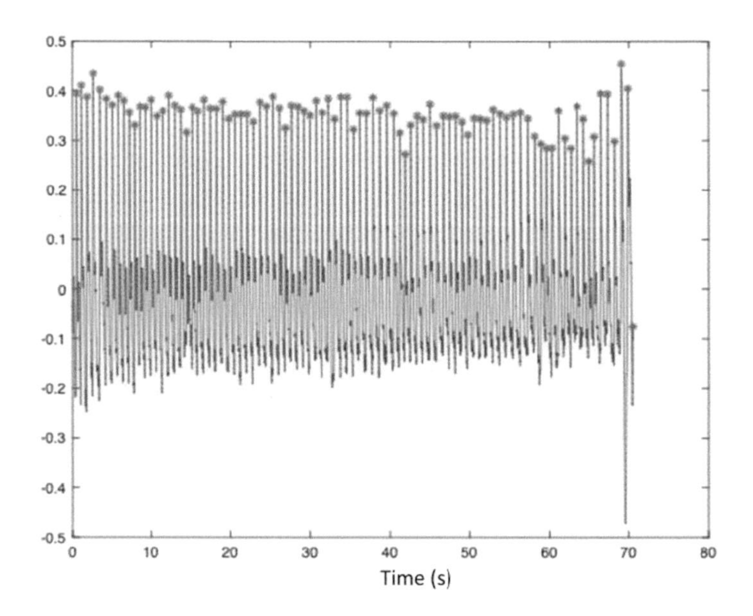

FIGURE 12.2 The ECG data collected during a Valsalva maneuver. The signal is band-pass filtered and the QRS peaks are marked. Notice a misdetection of a QRS peak at the end.

6. IBI Correction

Note that the IBI values are calculated only at the points when a heartbeat is detected. Therefore, the IBI measurements are not made uniformly in time because of the changing nature of the heart rate. However, in Digital Fourier Transformations, like FFT, the inherent assumption is that the digital data is sampled at uniformly spaced time points, i.e. the IBI measurements are taken for instance every second or so. In order to correct for this discrepancy, we need to do an interpolation and find the IBI measurements at uniformly taken time points. An alternative and probably easier method is that the IBI value can be kept at the originally sampled time points, and the interpolated IBI values can be inserted in between the original measurements of IBI so that original IBI measurements are separated by the correct time intervals, as shown in Figure 12.3. For instance,

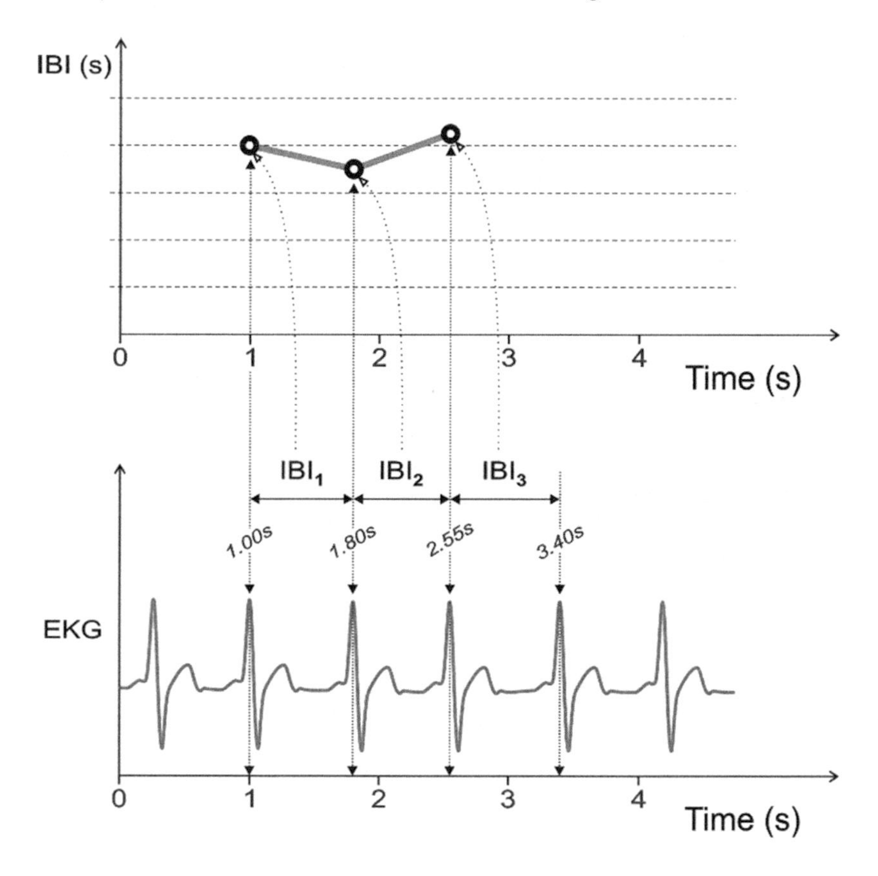

FIGURE 12.3 Finding the inter-beat intervals (IBI) from the ECG signal and interpolating them on a uniform time grid.

let us say that IBI values of 0.80 s, 0.75 s and 0.85s are found at time = 1.80 s, 2.55 s and 3.40 s, respectively. There are actually 749 sampling points missing between the first two points at the sampling rate of 1000 samples/s. A linear line is drawn from 0.8 to 0.75 and 749 intermediate IBI values are calculated on this linear interpolation line. Then a new array of IBI value can be made for the entire recording using these interpolated values inserted in between the original ones. Notice that the number of intermediate IBI values will be different for each consecutive IBI pair, depending on the IBI values. The length of the IBI array will increase approximately by a thousand times assuming an average heart rate of one pulse per second. A sample corrected plot of IBI values are shown Figures 12.4 and 12.5.

7. Taking the FFT

   When FFT is performed on this corrected IBI array (e.g., fft (IBI) as in Figures 12.6 and 12.7), there will be many coefficients produced by the FFT algorithm because the frequency axis will be ranging from 0 to 1000 Hz. But we are only interested in the 0–0.5 Hz range at the lower end of the spectrum. We will have to generate

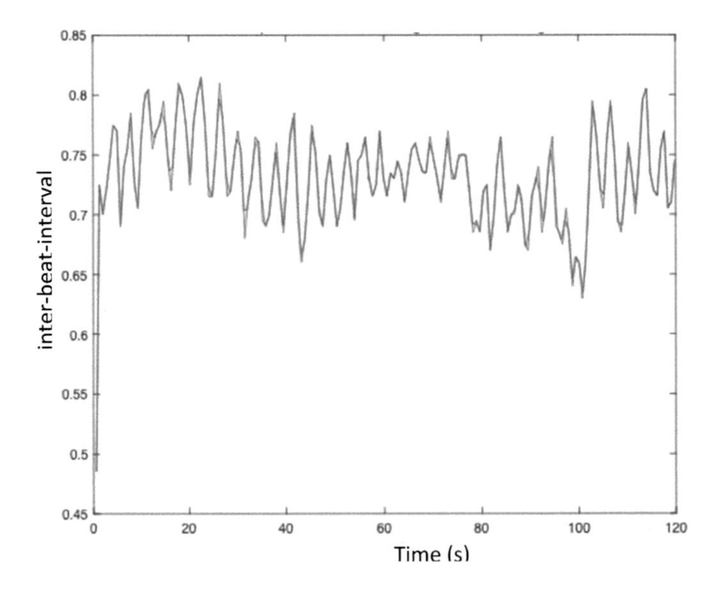

FIGURE 12.4 The inter-beat-intervals (IBI) found from the ECG signal in Figure 12.1. The IBI values are interpolated at 1 ms time intervals between the original measurements.

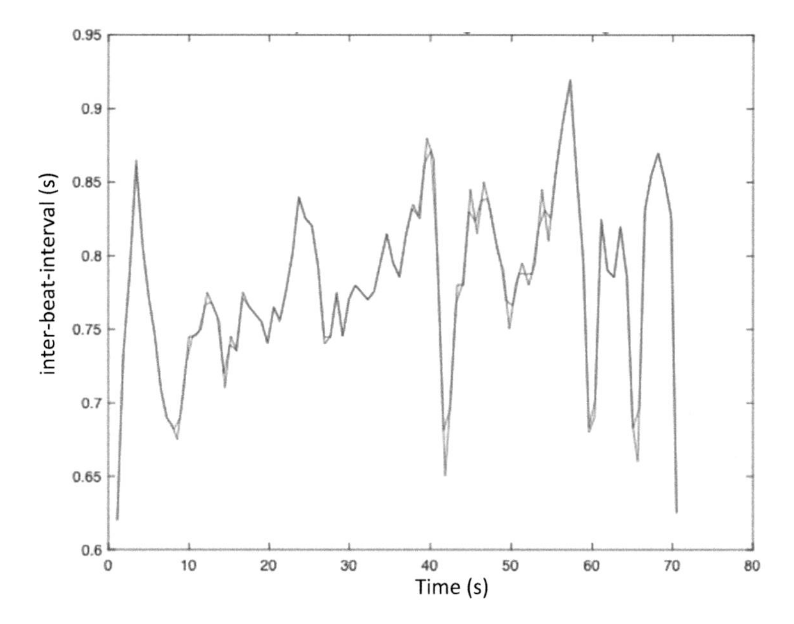

FIGURE 12.5 The inter-beat-intervals computed for the ECG data shown in Figure 12.2 during Valsalva maneuver. The IBI values are interpolated at 1 ms intervals. Notice the large but slow changes in the IBI.

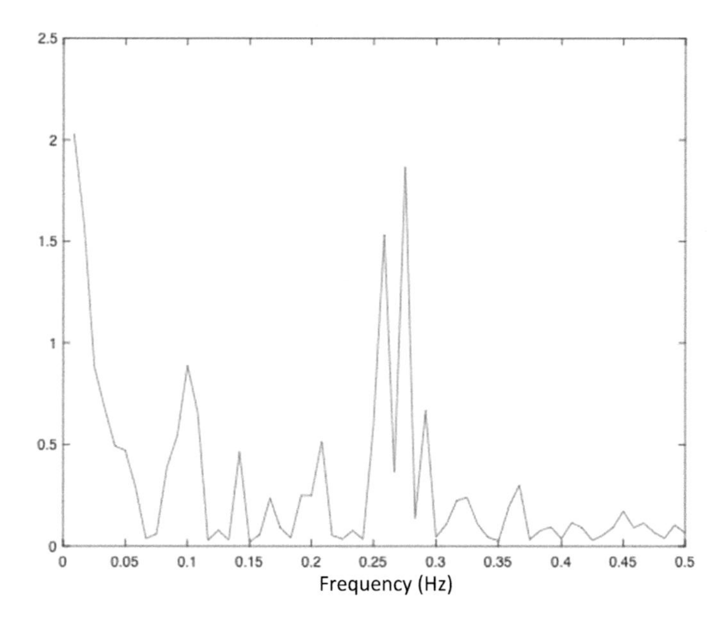

FIGURE 12.6 The FFT of the IBI data shown in Figure 12.4. The vertical axis is in arbitrary units. The LF/HF ratio is computed as 0.47.

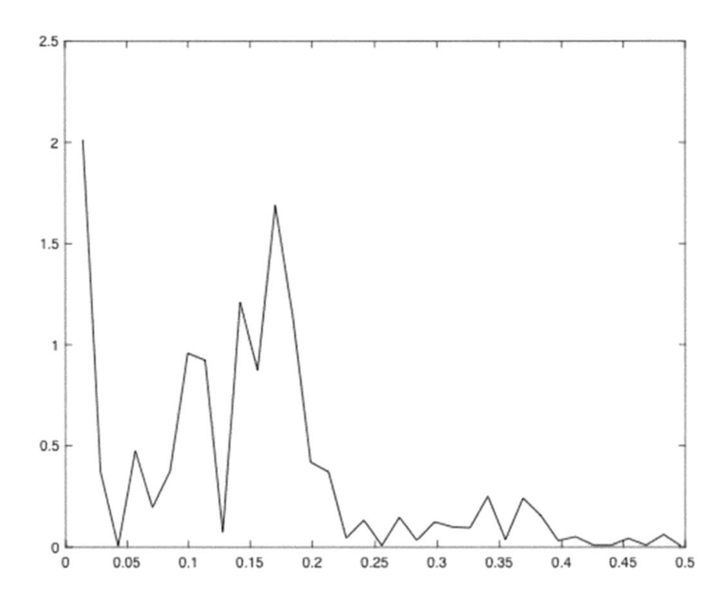

FIGURE 12.7 The FFF of the IBI shown in Figure 12.5. The low-frequency components (0.04–0.15 Hz) are much higher compared to the normal breathing in Figure 12.6. The LF/HF ratio is found to be 0.78, which implies a higher LF vs. HF power compared to normal breathing.

the frequency axis for the FFT plot using the correct sampling rate: x_axis = 0:1/L:fs-1/L, where fs =1000 samples/s and L = 100 s.

8. Computing LF and HF Power

   Each point in the FFT spectrum represents power within a frequency range of 1/L Hz (frequency resolution), where L is the total signal length in seconds (L = 100 s in our case which corresponds to 0.01 Hz frequency resolution). To find the power that falls into the LF or HF band, we need to find the FFT coefficients for those frequencies and take the sum of their squares, since the FFT coefficients are in units of $V/\sqrt{Hz}$ and the unit for power is V^2/Hz. The index of the coefficient that corresponds to an arbitrary f1, for instance, is f1/(1/L) + 1 or f1xL + 1. We are adding 1 to the index because the first coefficient represents the power at DC or zero Hz. More specifically, in this studio because we recorded for 100 s, the FFT coefficients for the LF band would have the index numbers of 0.04x100+1 to 0.15x100 + 1, or from fifth to sixteenth coefficients in the FFT array. Thus, the LF power will be calculated as:

$$\text{power} = \sum_{i=5}^{16} \text{abs}\left[\text{FFT}(i)\right]^2$$

where FFT(x) = fft(IBI);

9. After finding the HF power for the 0.15–0.4 Hz band following similar steps, the LF/HF ratio can be calculated. The LF and HF bands indicate relative contributions of sympathetic and parasympathetic control of the heart rate, respectively, as discussed in the Background section (S12.1) above.

10. Note that a sample MATLAB code for HRV analysis (HRV_analysis.m) is included in the enclosed CD along with samples of Resting and Valsalva maneuver data.

11. Finally compare your results with those outlined in the figures of this studio.

For your Lab Report:

In Results:

1. Include the raw and filtered ECG waveforms, the IBI plot, and the FFT spectrum.

2. Compare the resting HRV with the Valsalva HRV in terms of power in each frequency band or LF/HF ratio.

3. Include your MATLAB code in your report.

In Theory Section:

Talk about theory of Heart Rate Variability and how it is used clinically.

In Discussion:

1. Discuss your results with references from literature.

2. Discuss how HRV can be used in research.

REFERENCES AND MATERIALS FOR FURTHER READING

References marked with an asterisk (*) are recommended to those interested in expanding on the content of this chapter.

1. *Billman, G.E. Heart rate variability — a historical perspective, *Front. Physiol.* (2011) 2:86.
2. Bigger, J.T. Jr.; Fleiss, J.L.; Steinman, R.C.; Rolnitzky, L.M.; Kleiger, R.E.; Rottman, J.N. Frequency domain measures of the heart period variability and mortality after myocardial infarction, *Circulation* (1992) 85:164–171.
3. *Task Force of the European Society of Cardiology and the North American Society of Pacing and Electrophysiology. Heart rate variability: standards of measurement, physiological interpretation, and clinical use, *Circulation* (1996) 93:1043–1065.
4. *Neuman, M.R. Vital Signs: Heart Rate, *IEEE Pulse* (2010) 1(3):51–55.

Studio 13

AC Impedance of Electrode-Body Interface

S13.1 BACKGROUND

Electrodes are used for recording several different biological signals. These include ECG (EKG) electro-cardiograph, EMG; electro-myograph, EEG; electro-encephalograph and others. For these types of physiological signals skin surface electrodes, mostly disposable, are commonly used. These electrodes have significant contact impedance that can affect the quality of the recordings being made. In order to design biomedical equipment, the characteristics of the electrodes need to be well understood so that the amplifier design can be optimized for the best performance.

The single time constant model is shown in Figure 13.1. More complex models that include a constant phase element have also been developed. In general, the electrode impedance varies as a function of electrode current and frequency. The linear model shown above is an approximation that assumes that the electrode impedance does not change by electrode current, and the frequency dependency of the impedance is that of a first-order system, and thus it can be modeled with fixed electronic components. R1 represents the current pathway that is provided by the faradaic reactions at the electrode/electrolyte or electrode/tissue interface. C1 is the capacitance that occurs due to the accumulation of opposite charges at the interface between the solid phase (electrode metal) and the liquid phase (electrolyte) that cannot combine due to a chemical energy barrier. R2 is the resistivity of the volume of tissue between the signal source and the

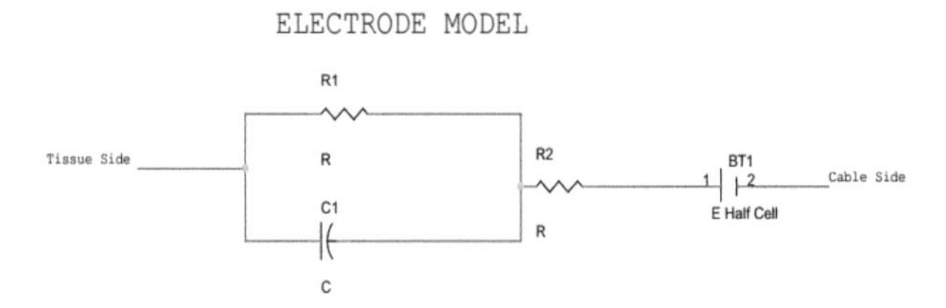

FIGURE 13.1 Single time constant electrode model.

electrode, such as the muscles and other tissues between the heart as the signal source and the recording electrodes on the skin. BT1 is the half-cell potential that develops at any metal electrode in physical contact with a solution of its ions (i.e., electrolyte).

S13.2 OVERVIEW OF THE EXPERIMENT

In this studio we will measure the series impedance of a pair of ECG electrodes on a human subject. This will be done by attaching a pair of electrodes on the arm of a subject and applying a small AC current from a signal generator to the electrodes while measuring the potential across the electrodes. The values of the two resistors and the capacitor in the model will be calculated in this experiment. The half-cell potential (BT1) is usually not much of an interest to the experimenter because AC amplifiers are used with most biological signals (except electro-oculogram – EOG) and the half-cell potential of an electrode varies a lot depending on the electrolyte gel and the temperature.

S13.3 LEARNING OBJECTIVES

The objectives of this studio are to:

- Understand the electrode equivalent circuit.

S13.4 SAFETY NOTES

It is extremely important that the subject is not connected directly to the ground of any of the instruments used during this studio – e.g., oscilloscope, voltage supply, computer. In hospitals the clinical equipment has isolated grounds that prevent the passage of AC or DC current from such equipment to the patients. Isolated grounds are a safety measure to protect patients in case of faulty equipment. In the worst situation, a direct connection between the subject and the ground of a faulty apparatus may result in a fatality. All electric equipment in the laboratory should be tested for safety prior to usage. This is normally the responsibility of the host institution.

S13.5 EQUIPMENT, TOOLS, ELECTRONIC COMPONENTS AND SOFTWARE

Additional information about the use of the items required in this studio can be found in Studio 1 and in the Appendices.

Equipment:

a. Breadboard.
b. Power supply (+/–12 V).
c. Multimeter.
d. Oscilloscope.
e. Signal generator
f. Data Acquisition Board (DAQ).
g. Computer.
h. Disposable ECG electrodes.

Tools:

a. BNC-to-micro-clip cables.
b. Alligator clip jumper wires.
c. Jumper wires for the breadboard.

Electronic Components (remember to download the available datasheets):

a. 1 Quad Operational Amplifier, TL084CP.
b. Various ¼ W resistors and capacitors.

   2–33K.

   2–1 K.

   4–1 M.

   2–0.1 μF.

c. Miscellaneous:

   3–2 pin header (optional).

   1–3 pin header (optional).

Software:

a. MATLAB®.

b. Software to interface your Data Acquisition Board.

S13.6 CIRCUIT OPERATION

The schematic for the circuit is shown in Figure 13.2. The signal generator is connector to JP1. The voltage divider networks, R1/R3 and R2/R4 protect the test subject from electrical shock and attenuate the signal from the generator. U1 is a gain of 33 differential amplifier where U1A and U1B are voltage buffers that drive U1C the difference amplifier. The output of the differential amplifier connects through JP2 to the data acquisition card. Power is provided at JP4.

S13.7 DETAILED EXPERIMENTAL PROCEDURE

1. Build the circuit given in the schematic on a breadboard.
2. Compare the circuit with the schematic and verify all the connections before applying the power supply.
3. Apply +/–12 volts DC to the breadboard. Apply a 5 Hz sinusoidal signal with ±0.1 V (peak-to-peak) amplitude from the signal generator to the input of the circuit (with respect to ground) to simulate an electrophysiological signal.
4. Now observe the output of the second Op-Amp on the oscilloscope (Channel 1) as well as the original sinusoidal signal from the signal generator (Channel 2) and compare if the waveforms are identical, except for amplitude, if there are any distortions in the output waveform, and if there is any phase shift.
5. Measure the gain of the circuit, and confirm that it agrees with the design.

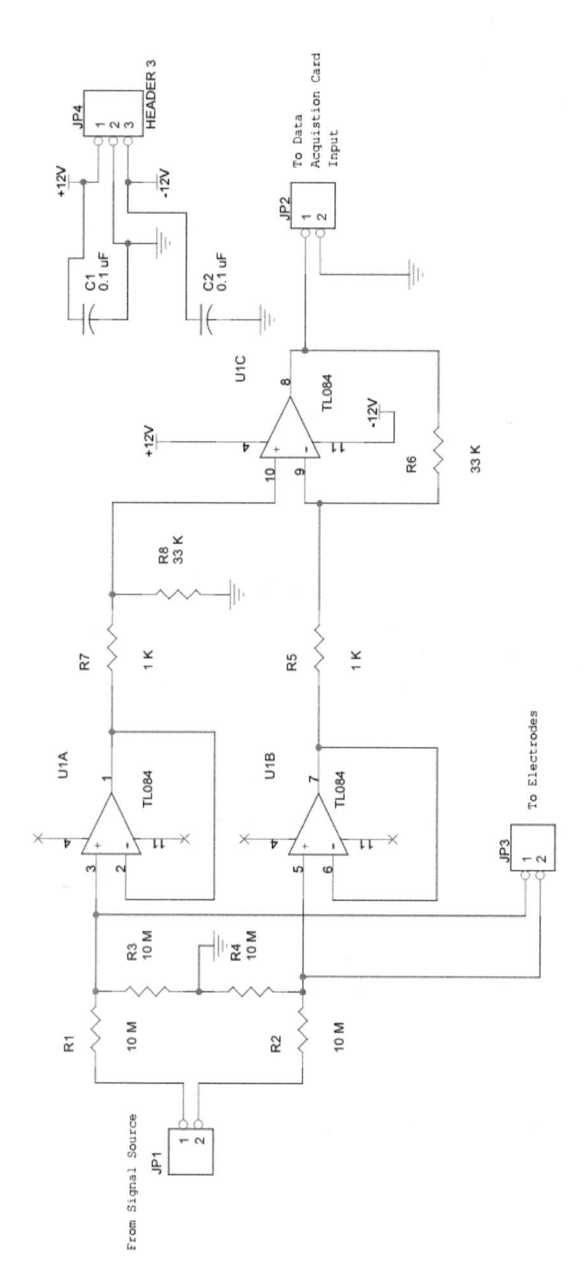

FIGURE 13.2 Schematic diagram electrode characterization amplifier.

6. Increase the frequency of the generator until the output signal drops by 3 dB, (the amplitude will be 70.7% of the amplitude at 5 Hz) making amplitude and phase shift measurements on the oscilloscope. This frequency should be approximately 100 KHz.

7. Connect JP3 to two disposable ECG electrodes on a subject's arm, one electrode near the inside of the wrist and one about 10 cm further up the arm using the alligator clip leads. Ensure that the Ag/AgCl parts of the electrodes are moist with electrolyte gel.

8. Return the signal generator to 5 Hz and set the output level to 3 volts peak to peak, note the new signal level at the output of the amplifier. Use these signal level to calculate the resistance of the two electrodes in series. Use the gain of the amplifier measured in step 5 and calculate the attenuation in the input network given the 2×10 MΩ resistors in series with the electrodes. Calculate the resistor value needed to provide this attenuation. Divide this number by 2 to obtain the total series resistance of each electrode.

9. Increase the frequency of the generator to again find the point at which the signal drops by 3 dB. Note the frequency.

10. Continue to increase the frequency until the signal level stops dropping. This should be between 20 and 50 KHz. Use this signal level to calculate the series impedance of both electrodes, and divide by 2 again to get the value for each electrode. Note: the signal will drop to between 10 and 40 mV peak to peak.

11. Subtract this value to get the value of the larger series resistor in the model, and with this value and the frequency found in step 9 calculate the capacitance of the electrode, using the formula for the corner frequency of a first-order R-C circuit.

$$V = I \times R$$

You now have determined the values of the three elements of the electrode model.

In your report, remember to include:

1. Introduction: Theory on simple electrode model.

2. Results: Show all the data and calculations.

3. Discussion: Discuss the problems encountered and how they were solved in building and testing of the circuit. Discuss the limitations of the electrode model. Discuss limitations of the circuit in terms of frequency and signal amplitude. Discuss why the excitation current is set at a low level.

S13.8 CIRCUIT TESTING AND TROUBLESHOOTING

1. Before turning the power on, carefully check if the voltage supply is applied to the circuit with the correct polarity. Using a voltmeter one lead connected to the ground, check the +/–12V and ground voltages at the power terminals of all chips. It is a time saving practice to do this as the first step of troubleshooting in any circuit, analog or digital.
2. Another good practice is to measure the total current that the circuit is drawing from the power supply (most modern power supplies have a voltage and a current display). If the current is higher than a few tens of a mA, or one of the chips is becoming too warm (be careful not to burn your fingers) you may conclude a broken chip or a short circuit due to an incorrect connection.

S13.9 QUESTIONS FOR BRAINSTORMING

a. What happens if you increase the value of R6?

b. What is the function of R7 and R8?

c. How could you characterize the electrodes with a two-time constant model, rather than then one time constant model used?

S13.10 IMPORTANT TOPICS TO INCLUDE IN LAB REPORT

a. The need for an electrode model.

b. Any compensation that maybe needed in the design of the physiological amplifier due to the electrodes characteristics.

c. Any problems that high electrode impedance may cause in an ECG or EMG.

REFERENCES AND MATERIALS FOR FURTHER READING

References marked with an asterisk (*) are recommended to those interested in expanding on the content of this chapter.

1. *TL08xx Datasheet*, Texas Instruments, SLOS081, May 2015, Dallas, TX.
2. *Assambo, C.; Baba, A.; Dozio, R.; Burke, M.J. Determination of the parameters of the skin-electrode impedance model for ECG measurement, *Proceedings of the 6th WSEAS Int. Conf. on Electronics, Hardware, Wireless and Optical Communications*, Corfu Island, Greece, February 16–19, 2007, pp. 90–95.
3. Neuman, Michael R. Biopotential electrodes. Chapter 5 in *Medical Instrumentation Application and Design*, Webster, J.G., 4th ed. John Wiley & Sons, Inc., New York.

Appendix I

Using Electronic Components and Circuit Design

A1.1 RESISTORS

Resistors are devices that impede the flow of electrons. They are linear devices that follow Ohm's Law, that is

$$V = I \times R$$

where:

V = voltage in volts
I = current in amperes
R = resistance in ohms.

There are many types of resistors. Two broad classes are fixed and variable resistors

A1.1.1 Fixed Resistors

This is the most common type of resistor; it has a specific resistance value.

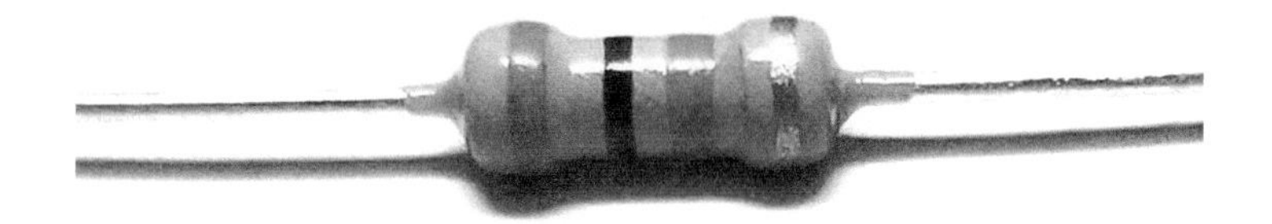

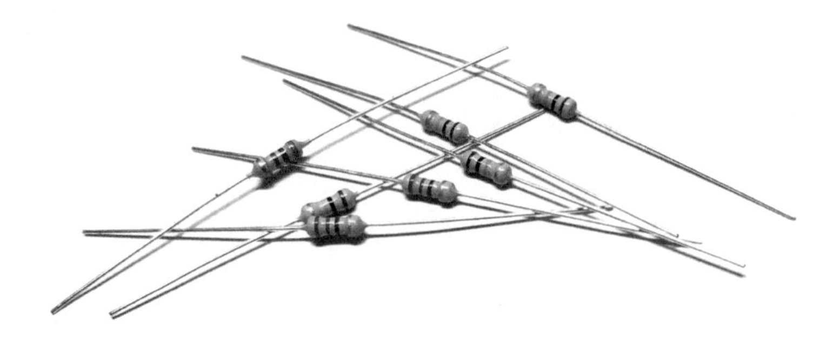

A1.1.2 Variable Resistors

This is a device that consists of a fixed resistor element and a slider that moves across the element. The device has three (3) connections, two for the ends of the fixed resistor and one (1) for the wiper. It can be wired as a voltage divider, with a voltage present across the ends of the fixed element, and the output selected by the wiper, or as a variable resistor, with one end of the fixed element and the wiper being used to create an adjustable resistor.

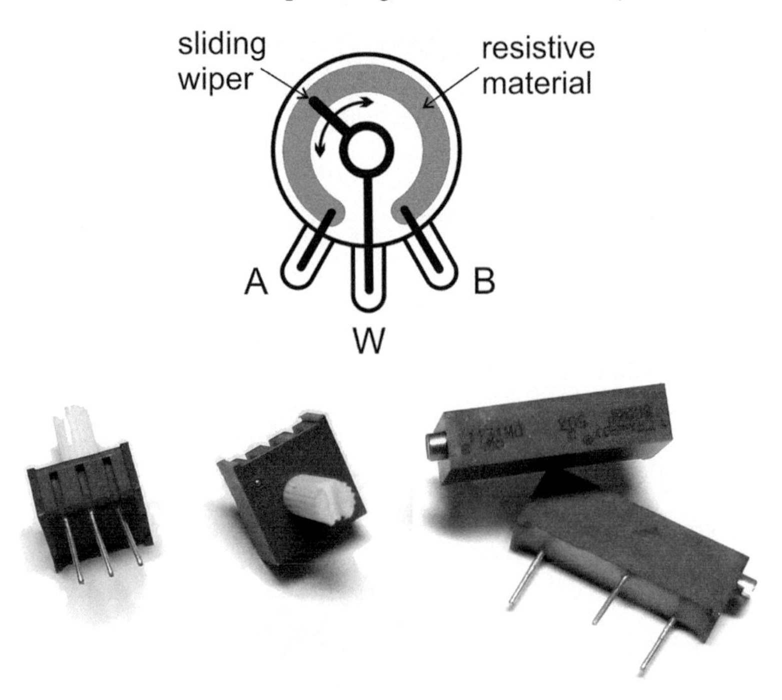

A1.1.3 Fixed Resistor Types

- Carbon Composition: These used to be the most common type, but today they are not used as much, mostly due to the loss of their cost advantage over other types, and due to the proliferation of surface mount components. It is made by placing carbon granules that fit between to wires (component leads) and are encapsulated.

- Carbon Film: To make this, a carbon ceramic film is deposited on a form, and the film is cut into a helix. The turns in the helix determine the resistance, as does the length, diameter and thickness of the film.
- Metal Oxide Film: This is the most common type of resistor today. These resistors are very temperature stable and can easily be made to tight tolerances, 1% being the most common. It is made in a similar manner to the carbon film, but has much better temperature coefficients.
- Metal Film: These are very similar to Metal Oxide film resistors
- Thin Film: This is the most common type of surface mount resistor in use today. It consists of a ceramic substrate with a metal oxide film deposited on it.
- Wire Wound: These are used mostly in power applications where many watts of dissipation are required. A wire is wound on a ceramic tube and covered with an enamel.

A1.1.4 Resistor Color Codes

Carbon composition, carbon film and metal oxide film resistors are usually marked with colored bands to indicate the resistance. The table below is the key.

Color	First significant figure	Second significant figure	Third significant figure	Multi-plier	Tolerance %	Temp Coeff. (ppm/°K)
Black	0	0	0	1		250
Brown	1	1	1	10	1	100
Red	2	2	2	100	2	50
Orange	3	3	3	1 K		15
Yellow	4	4	4	10 K		25
Green	5	5	5	100 K	0.5	20
Blue	6	6	6	1 M	0.25	10
Violet	7	7	7	10 M	0.1	5
Grey	8	8	8	100 M	0.05	1
White	9	9	9	1 G		
Gold				0.1	5	
Silver				0.01	10	

Resistors with three or four bands have two significant figures followed by the multiplier. The multiplier is the number of zeros to be padded to the right of the first two numbers. If it only has three bands, it is 20% tolerance, otherwise the fourth band is the tolerance. If it has more than four bands, it has three significant figures. Some examples are:

- Orange, Orange, Orange is 33 K 20% tolerance.
- Orange, Orange, Orange Gold is 33 K 5% tolerance.
- Orange, Orange, Brown, Red, Brown is 33.1 K 1% tolerance.
- Orange, Orange, Brown, Red, Brown, Brown is 33.1 K 1% tolerance 100 ppm.

Note: there will be a gap between the last significant figure and the tolerance in the bands, so you can determine which side of the resistor the code starts from. It is a good idea to verify the resistance with your Ohmmeter. The failure rate code is not in common usage.

The resistors are manufactured in standard values since producing every single possible value would not be practical. These preset values are known as the E3, E6, E12, E24, E48, E96 and E192 series, where the number indicates the number of standard resistor values in each decade. For instance, in E12 series there are 12 standard values between 1 and 10 Ohms: 1, 1.2, 1.5, 1.8, 2.2, 2.7, 3.3, 3.9, 4.7, 5.6, 6.8, 8.2 and 10 Ohms. This pattern repeats itself in the next and following decades by multiplying the standard values by 10, 100, etc. That is, 10, 12, 15 ….. Ohms in the next decade. In higher series, there is more options in between these standard values while keeping the values in the lower series. This provides some convenience since the engineer knows beforehand what values are available while designing a circuit.

A1.2 CAPACITORS

There are many different types of capacitors mostly defined by the dielectric material used between the plates. They all conform to the same basic laws and are constructed of two conductive plates separated by an insulating dielectric material. The differences in construction of the different types of capacitors lead to different properties that the students must be aware of for some demanding applications, particularly regarding the leakage currents, operating frequency limitations, tolerance values, and polarization requirements. In general, the physical size of the capacitor increases with increasing capacitance and the voltage rating. However some novel technologies allow drastic reduction in the size of the capacitors. We will discuss the different types.

- Ceramic.

  These are very common and used in most frequency ranges, in applications from audio to microwaves, power supplies, etc. They normally range in values from picofarads (pF) to several microfarads (μF), depending on the operating voltage. They are available in many different physical configurations. ceramic capacitors are widely used in surface mounted (SMD) applications.

- Aluminum electrolytic capacitor.

  These are polarized devices, that is they must have a DC voltage across them to operate, and the correct polarity must be observed. They have higher capacitance then non-polarized devices for the same physical size, such as ceramic capacitors, but generally they cannot be used at frequencies above ~100 kHz where they become inductive.

- Tantalum capacitor.

  This is a type of electrolytic capacitor that has better high frequency response then the aluminum electrolytic, and offer higher densities, i.e., smaller size for the same capacitance. They are more expensive then aluminum electrolytics.

- Silver mica.

  These capacitors were developed for radio frequency (RF) applications. They offer high stability and low loss and generally have values less than 1000 pF. They are not available in surface mount, and have therefore not been widely used in recent years.

- Polystyrene film.

  These are inexpensive capacitors with good tolerances, are limited to frequencies below 1 MHz, and are only readily available as leaded parts today.

- Polyphenylene sulfide (PPS).

  These film capacitors are available in surface mount, and have higher tolerances, +/−5% and +/−2%. They have low temperature coefficients which makes them ideal for tight tolerance applications. They are generally useable in applications under 1 MHz.

- Supercap.

  These polarized capacitors offer very high capacitance values, up to thousands of farads, but generally have low operating voltages, below 5 volts.

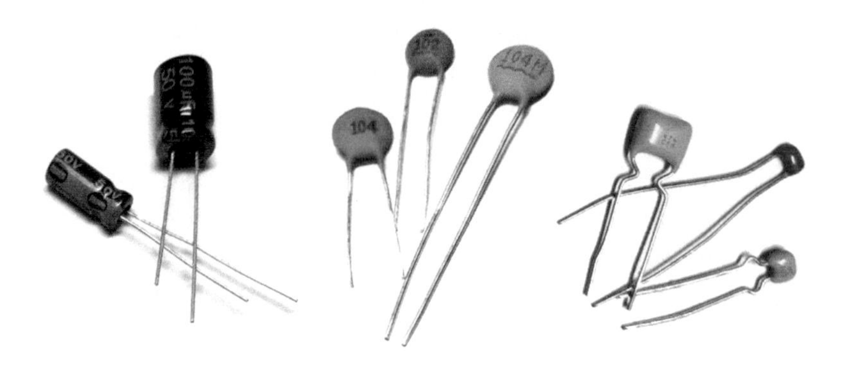

A1.2.1 Marking of Capacitor Value

Physically large capacitors may have the values clearly written on them. For smaller parts, they are marked numerically. In the past, some used color codes similar to the resistor color coding scheme. However this is not used today. There is no industry standard in marking capacitors, so if in doubt, check the manufacturer's data sheet. Many capacitors are marked with a three digit code, the first two digits being the capacitance and the third digit a multiplier, ten to the power indicated by the number. A capacitor marked as 104 would be 100,000 picofarads, or 0.1 μF. Another style is the decimal point format, e.g., "0.01" which is followed by a letter that indicates tolerance. The unit for this format is always μF, even though it is usually not indicated. Capacitors may also be marked with a tolerance they follow the EIA tolerance marking code.

Letter	Tolerance
Z	+80%, −20%
M	+/−20%
K	+/−10%
J	+/−5%
G	+/−2%
F	+/−1%
D	+/−0.5%
C	+/−0.25%
B	+/−0.1%

The working voltage is usually marked directly on the capacitor. A capacitor marked 6 V would have a working voltage of six volts. The working voltage is the maximum voltage the capacitor can work at for its rated life. Derating the voltage is a good idea for high reliability. So, if you have a circuit running at +6 V, you would select a capacitor with a higher rating, 9 V or higher. Polarized capacitors are marked with a polarity indicator. Aluminum electrolytics usually have a black stripe indicating the negative terminal. Surface mount capacitors will have a stripe on the positive terminal.

A1.3 OPERATIONAL AMPLIFIERS

Operational amplifiers (Op-Amps) are differential input, single ended output devices. Ideal Op-Amps have attributes as follows:

- Infinite differential gain.
- Zero common mode gain.
- Zero input bias current.
- Zero input offset voltage.
- Infinite input impedance.
- Zero common mode gain.
- Zero output impedance.
- Infinite bandwidth.
- Infinite slew rate.

- Differential gain is the gain between the + and – inputs to the output.
- Common mode gain is the gain when the + and – inputs are tied together, and the same signal is applied to both inputs.
- Bias current is the current that flows into or out of the input pins.
- Offset voltage is a small voltage that when applied between the two input pins that makes the output voltage zero.
- Input impedance is the load that the driving circuit sees at the input to the Op-Amp.
- Output impedance is the impedance in series with the output connection internal to the Op-Amp.
- Bandwidth is the highest frequency the Op-Amp can operate at. This is gain dependent and formulated such that the product of Bandwidth × Gain is a constant.
- Slew rate is how fast the output moves, usually measured in volts/μsec.

Op-Amp Symbol – + and – are inputs, vertical lines are power supply connections.

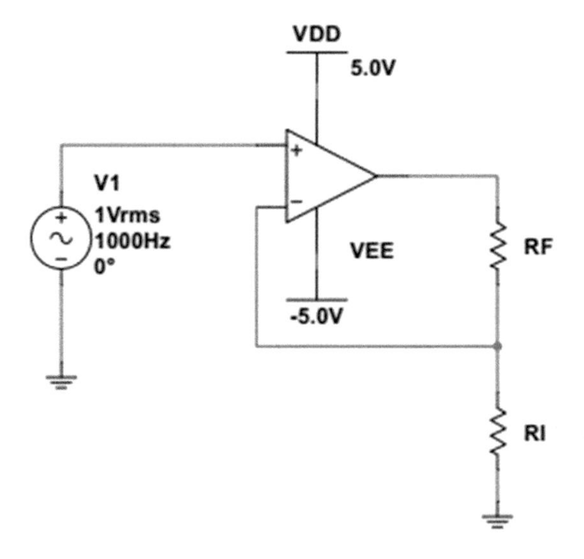

For a non-inverting configuration Op-Amp, the gain of the circuit can be defined as:

$$G = \frac{V_{out}}{V_{in}}$$

The resistors set the gain of this circuit which becomes:

$$G = 1 + \frac{RF}{RI}$$

Where RF is the feedback resistor and RI is the input resistor.

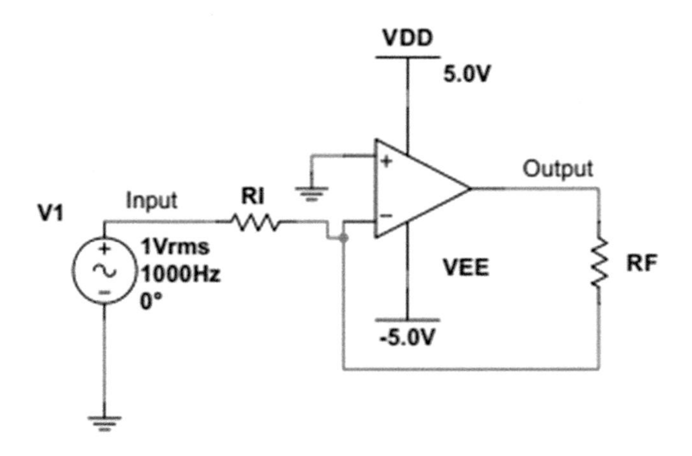

For the inverting configuration, the gain is:

$$G = -\frac{RF}{RI}$$

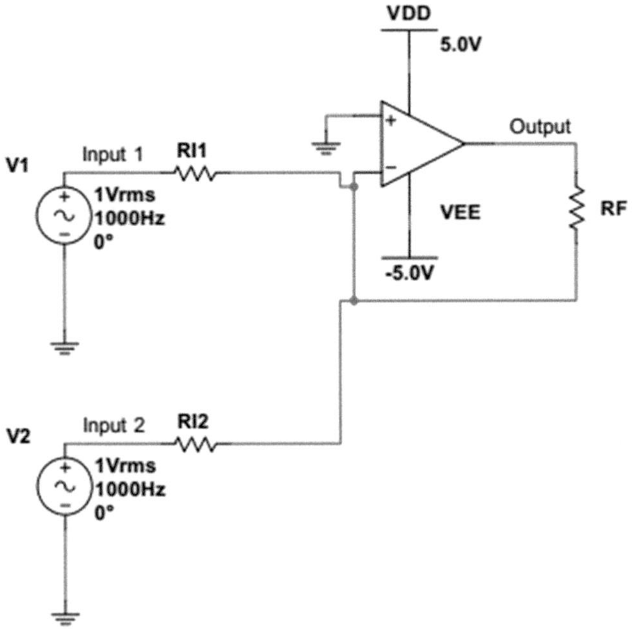

A summing amplifier has multiple inputs, with no theoretical limit. In the above case, the gain is the same from each of the two inputs as the formula in the prior example. For the gain from input 1 the denominator is RI1, and for the gain from input 2 has RI2 in the denominator.

A1.4 DESIGN ISSUES

When designing a circuit, the Op-Amp cannot just be selected at random, there are many issues to consider. The issues we will discuss will be sufficient for most situations, but it will not cover all cases.

A1.5 SLEW RATE

The slew rate tells us how fast an Op-Amp's output voltage can change with time. If we need to have a 1 MHz 10 volt p-p sine wave signal as our output, then we need to calculate the required slew rate. The formula is:

$$SR = \pi \times FBW \times Vpp$$

Where FBW is full power bandwidth, or 1 MHz and Vpp is 10 volts peak to peak. Then the minimum SR required in this application is 31.4 volts/μsec. On the manufacturer's datasheet, you need to find the minimum slew rate specification for the device at the power supply voltage you plan on using in your design. The slew rate will be specified for more than one power supply voltage in most cases. You should never use typical values when selecting a part, since only the minimum is guaranteed. You should always consider the worst case situation.

A1.6 VOLTAGE SWING

In the above example, we see we need to swing 10 volts peak-to-peak at the output. We need to ensure that our Op-Amp can do this. Again, we need to look at the datasheet for the output voltage swing, which is how close the output voltage can get to the power supply voltage. We need to check this at both high and low values at the current we will be delivering to our load. Depending on the Op-Amp we are looking at this voltage can be as close as 50 millivolts to the power supply voltage or a couple volts shy of it. A device where the output voltage can come close to the power supply voltage is called a rail-to-rail output. If we find that the Op-Amp can get to 1.0 volts shy from either the positive or negative supply, then we would need to have at least 12 volts between the Vplus and Vminus supply pins to ensure that we can swing 10 volts at the output.

A1.7 POWER SUPPLY

We have determined that we will use a nominal 12 volt supply. We now need to decide whether we will use +/−6 or +12 volts. If our signal can be AC coupled, that is we use a capacitor between the stages to take out the DC signal, then we can use either power supply configuration. If we need to amplify the DC component of the input signal as well as the AC, that is we need frequency response down to zero, we cannot use a capacitor in series in the signal pathway. If we need DC coupling and the output needs to be centered around 0 volts, or ground, then we must have +/− supplies. Otherwise we can use either configuration. However, we may need to design the input of the amplifier differently for each configuration. For the dual supply, there is nothing special to consider. You should have no issues with a ground referenced input. However, for the single supply configuration, more care must be taken. You need to find the input voltage range specification. Some devices have rail to rail input, and some even have inputs that can operate below the negative supply rail, then you can use the Op-Amp with a single supply rail. However, if the Op-Amp you have chosen requires the input voltage to be more positive then the negative supply voltage, which is ground in the single supply configuration, you must use a dual supply configuration for that device. For non-inverting configurations that have an input voltage below ground, in almost all cases a dual supply configuration will be required, since the input voltage at the Op-Amp pin will be going negative, and unless the signal input amplitude is very small, will most likely exceed the low input operating limit of the device.

A1.8 GAIN BANDWIDTH PRODUCT

The gain bandwidth product, or GBW, will tell us the highest frequency the Op-Amp can work at for a specific gain. If we design our amplifier to have a gain of ten, then for the 1 MHz signal, we need a GBW of at least 10 MHz.

A1.9 SELECTING RESISTOR VALUES

For the inverting amplifier with a gain of ten, we know the feedback resistor must then be ten times the value of the input resistor. How should we select this value? From the point of view of power dissipation, we want to make the resistance as large as possible, but there are other considerations that limit the maximum value. The two main issues are frequency response, and bias and offset currents. The circuit construction will add stray capacitance in parallel to the feedback resistor, creating a classical single pole low pass filter. Through hole parts will have a much higher capacitance then surface mounted parts. It is not straight forward to calculate this capacitance since it depends greatly on the circuit board layout that it is built with. Generally speaking, we can conservatively select a maximum resistor value for the feedback resistor if we use the following formula:

$$R_{max} = 10^{10} / F_{max}$$

Where F_{max} is the highest frequency we want to amplify. So, for our 1 MHz amplifier it is safe to use a 10 K feedback resistor from a frequency response point of view. The bias current is the current that the amplifier's input can draw or source. This creates a DC error in the output voltage. In most cases this is small and can be ignored. However, it should always be confirmed that it can be ignored. So, for our gain of ten amplifiers, our input resistor would need to be 1 K. The value of both these resistors in parallel (in the Thevenin equivalent circuit) is then 909 ohms. If we can tolerate a 10 mV shift in output voltage in our application, then we can calculate the maximum offset current we can accept at 10 mV/(909 × 10) = 1.1 μA. Otherwise we need to either reduce the resistor values to be within the specification, select a different Op-Amp, or consider if the specification can be relaxed to comply. The last thing we need to consider is noise. The input resistor generates Johnson noise that is amplified by the Op-Amp. Johnson noise can be calculated with the following formula:

$$V_n = \sqrt{4k_bTBR}$$

Where: k_b is Boltzman's constant 1.38×10^{-23}
T is temperature in degrees Kelvin
R is the resistance in Ohms.
B is the bandwidth in Hz.

For our amplifier, we picked a device with a GBW of 100 MHz, so our bandwidth will be 100 MHz/10 or 10 MHz. Using a T of 300°K we get a noise voltage of 0.4 μV rms which is very small and can be ignored.

A1.10 OFFSET VOLTAGE

We should consider offset voltage as well. If we look at the datasheet for the device you selected for this design, we will find a specification for Input Offset Voltage. A typical amplifier might be 500 μV. If we multiply that by our gain of 10, we get 5 mV at the output of the amplifier. This 5 mV output offset will be in addition to the 10 mV offset we calculated above due to the input bias current.

A1.11 TROUBLESHOOTING THE CIRCUIT

Sometimes after assembling a circuit it does not function properly or at all. In these cases it is important to be able to determine what is wrong and fix the problem. Usually the problem is an assembly error. So the first step is to double check your work and make sure the circuit assembled matches the schematic. Common errors of this type are using a wrong component; you may have misread the color code of a resistor, for example, and not checked it first with an ohmmeter before putting it into the board. Another common problem is plugging a lead into the wrong column on the breadboard either to the left or the right of the desired one.

Sometimes the power is not wired up correctly, or not getting to the voltage input pins of the IC. This can be checked with a voltmeter directly at the ICs power pins. Sometimes the component used is defective. Components like diodes and transistors can be removed from the circuit and checked with an ohmmeter. A diode will exhibit a low resistance in one direction, and a high resistance when the leads are reversed. Make sure you plugged the diode in correctly where the line on the diode package indicates the cathode connection (usually the side with the straight line in the symbol).

Although we do not use any discrete transistors in this book, they can also be easily tested with a multimeter by checking the emitter-base and base-collector junctions in a similar fashion, or if your meter has one, the h_{fe} test will indicate if the transistor's gain is high enough.

Integrated circuits (ICs) are more difficult to test. If you find nothing else wrong, you can substitute a different one and see if it corrects the problem. With the IC, first make sure you located pin #1 properly.

Capacitors can be challenging to test as well. If your meter can measure capacitance, that is the best method. Otherwise, if the capacitor value is large, i.e., in the microfarad range, it is usually possible to see the capacitor charge up when connected to the ohmmeter leads. The resistance initially is low, and then increases to that of an open circuit. Use a 200 kΩ scale to do this test. With electrolytic capacitors, make sure the polarity is correct, the black stripe on the capacitor indicates the negative lead.

Appendix II

Required Equipment and Materials

This section is primarily for the instructor using this book. We have made a consolidated list of all the equipment and material used in the studios to make it easier to procure the required material. We have provided manufacture's part numbers. These are only a guide and substitution can be made to both the manufacturer and the value of component or type of equipment referenced. Many of the components specified are not critical and can vary significantly without having an effect on performance. It is up to the instructor to use discretion in making these changes. In the United States, the electronic components are readily available from suppliers such as Digi-Key and Mouser. Components such as resistors and capacitors can be purchased inexpensively in quantities of 100 or 200 pieces in many cases, which is the most economical way to purchase these types of parts. We have put connectors on the parts list to easily attach the cables to the test circuit. However, in most cases, these can be omitted, and leads can be directly attached to wires pushed into the protoboard or attached to component leads. The only connectors required are those that fit the connectors that come with commercial transducers. Even this can be eliminated if the connector on the transducer cable is cut off and the wires connected directly to the circuit.

A2.1 COMPUTER SOFTWARE

Many of the studios use MATLAB®. We have used the softscope tool available in the 2013 and earlier editions. We have also used the Analog Input Recorder in the data acquisition toolbox available in the 2014 and later additions of MATLAB. Other software may be substituted, but we are providing sample

MATLAB code to use in many of the studios. You can use any PC that meets the requirements for the version of MATLAB you use. In addition, you need to provide a data acquisition unit to collect data. The National Instruments USB-6000 series are sufficient for our purposes but any MATLAB compatible analog data acquisition card that handles DC inputs should work.

A2.2 TEST EQUIPMENT

- Digital 2-channel oscilloscope 30 MHz minimum.
- Oscilloscope Probes.
- 1 MHz or better Function generator.
- Digital multimeter with Volts, Ohms and Amperes.
- Dual power supply adjustable from 0 to 15 v, 100 mA minimum.
- Protoboard (sometimes called breadboard).

A2.3 RESISTORS

The following are 5% ¼ watt parts:

VALUE	Part Number
270	CF14JT270R
470	CF14JT470R
1 K	CF14JT1K0
2.2 K	CF14JT2K20
2.7 K	CF14JT2K70
4.7 K	CF14JT4K7
6.8 K	CF14JT6K8
10 K	CF14JT10K0
12 K	CF14JT12K0
20 K	CF14JT20K0
47 K	CF14JT47K0
68 K	CF14JT68K0
91 K	CF14JT91K0
100 K	CF14JT100K
150 K	CF14JT150K
200 K	CF14JT200K
820 K	CF14JT820K
1 M	CF14JT1M0
3.3 M	CF14JT3M30
6.8 M	CF14JT6M80

The following are 1% ¼ watt parts:

VALUE	Part Number
51 Ω	RNF14FTD510R
68.1 Ω	RNF14FTD68R1
100 Ω	RNF14FTD100R
332 Ω	RNF14FTD332R
1 K	RNF14FTD1K00
10 K	RNF14FTD10K0
22.1 K	RNF14FTD22K1
49.9 K	RNF14FTD49K9
1 M	RNF14FTD1M00

The following are trimpots (may be called variable resistor or potentiometer):

VALUE	Part Number
10 K	PV36W103C01B00 Bourns
20 K	PV36W203C01B00 Bourns
50 K	CT6EP503 Copal

A2.4 CAPACITORS

All capacitors are 10% 50 V or higher ceramic unless noted.

VALUE	Part Number	Type	Note
100 pF	K101K10X7RF5UH5	X7R	
0.001 μF	K104K10X7RF5UH5	X7R	
0.0027 μF	FG28C0G1H272JNT06	COG	5% 60 Hz countries
0.0033 μF	FG28C0G1H332JNT06	COG	5% 50 Hz countries
0.01 μF	RDER73A103K3M1H03A	X7R	
0.047 μF	K473K15X7RF5TH5	X7R	
0.1 μF	K104K10X7RF5UH5	X7R	
0.15 μF	B32529C0154J289	Film	5%
0.33 μF	B32529C1334J189	Film	5%
1 μF	C330C105J5R5TA	X7R	5%
2.2 μF	TAP225K016SRW	TANT	10% 16 V

A2.5 SEMICONDUCTORS

Part Number	Description
TL084CP	Quad Op-Amp
TL082ACP	Dual Op-Amp, can use TL084 instead
1N4148	General purpose diode
AD7575JNZ	Analog to digital converter
CD4511BE	7-segment LED driver
HDSP-513A	7-segment display
SN74HC14N	Hex Inverter (Schmitt trigger)
1NA126P	Instrumentation amp
LM833NG	Audio Op-Amp
LM335Z	Temperature sensor
LM358AP	Dual Op-Amp

A2.6 MISCELLANEOUS COMPONENTS

Part Number	Description	Manufacturer
B57164K0103J000	Thermistor	EPCOS (TDK)
IA1205S	Dual power supply	XPPower
IL300-G	Opto-coupler	Vishay
Fort 1000	Force transducer	World Precision Instruments
AOM-4546P-R	Electret microphone	PUI Audio

A2.7 MISCELLANEOUS ITEMS

- BNC-to-microclip cables.
- Banana-to-banana cables.
- Banana plug to microclip cables.
- Jumper wires for the breadboard.
- AC Physiological Amplifier (e.g., Grass IP511 or equivalent).
- 3-lead cable for the Amplifier.
- EMG or ECG electrodes.
- Conductive ECG electrode gel (optional).
- Tubing 2” 5/16” ID latex rubber McMaster 5234K34.

- 9-volt battery and connector.
- Soldering iron.
- Solder.

Alligator clips.

Banana plug.

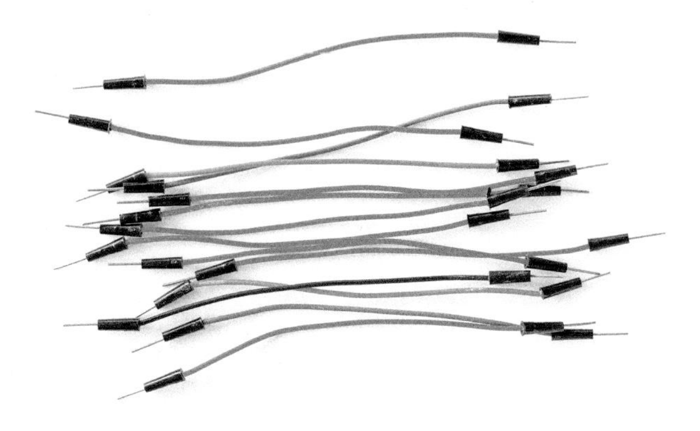

Jumper wires for breadboard.

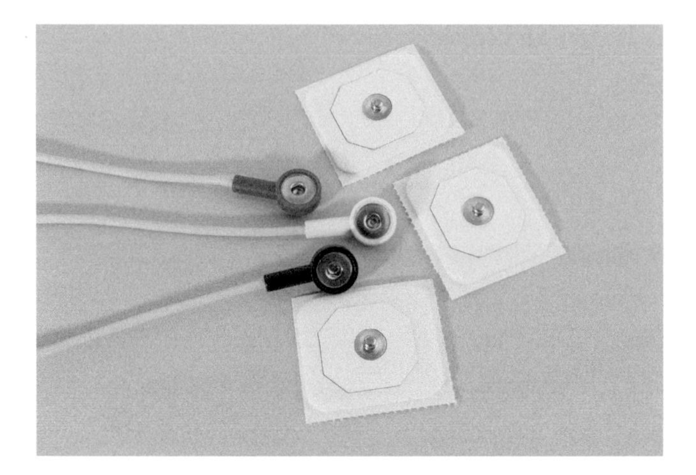

Disposable ECG electrodes.

BNC cable.

Index